GLOBAL CLASSIC LANDSCAPE DESIGN EXPLORATION HIGHLIGHTS

Global Classic Landscape Design Exploration Highlights

全球经典景观设计探索集锦 II

《景观设计》杂志社 编

大连理工大学出版社

图书在版编目（CIP）数据

全球经典景观设计探索集锦：全4册 /《景观设计》
杂志社编. -- 大连：大连理工大学出版社, 2011.9
ISBN 978-7-5611-6520-1

Ⅰ. ①全… Ⅱ. ①景… Ⅲ. ①景观设计—作品集—世
界—现代 Ⅳ. ①TU-856

中国版本图书馆CIP数据核字(2011)第182901号

出版发行：大连理工大学出版社
（地址：大连市软件园路80号 邮编：116023）
印　　刷：利丰雅高印刷（深圳）有限公司
幅面尺寸：245mm×245mm
印　　张：60
字　　数：1300千字
出版时间：2011年9月第1版
印刷时间：2011年9月第1次印刷
策划编辑：苗慧珠
责任编辑：刘晓晶
责任校对：万莉立
版式设计：王　江　赵安康　张建实

ISBN 978-7-5611-6520-1
定　价：880.00元（全4册）

电　话：0411-84708842
传　真：0411-84701466
邮　购：0411-84708943
E-mail:dutp@dutp.cn
http://www.landscapedesign.net.cn

目录 Contents

广场 _ Square

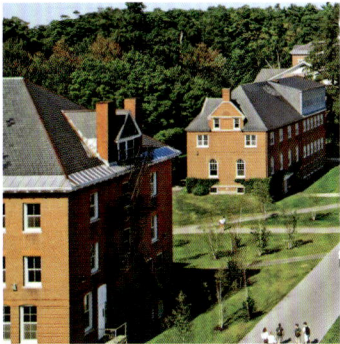

校园 _ Campus

目录 Contents

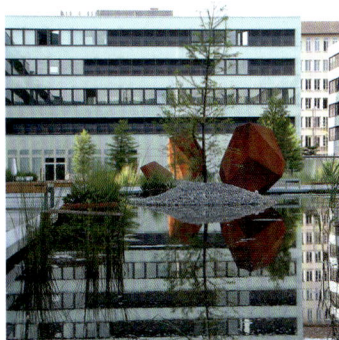

广场

善牧者休闲广场

Paseo del Buen Pastor

撰文：Jimena Martignoni　　图片提供：Sandra Siviero　　翻译：张璐

1　开放空间中心和绿色砖地
2　原监狱所在地现在是综合大楼和咖啡厅前的通道

象征科尔多瓦五条河流的喷泉上层水流

善牧者休闲广场的西班牙语名为 "Paseo del Buen Pastor" 是阿根廷科尔多瓦省首府科尔多瓦市新近开放的一处集商业文化为一体的多功能公共空间。科尔多瓦省地处阿根廷中部，有五条河流流经该省，构成了当地重要的水文系统。该水文系统融合了开放空间的布局理念，当地的景观设计也大多以此作为参照。

该项目于 2007 年竣工，颇具争议性，因为是将女子监狱和年代久远的教堂原址进行重建。教堂始建于20 世纪初，归善牧修女教会所有，为典型的西班牙殖民时期建筑风格，拥有开阔的走廊和露台；女子监狱毗邻教堂而建，高大而坚固的围墙与周围环境格格不入，给人一种冷冰冰的感觉。

女子监狱和教堂建在科尔多瓦市最繁华的广场上，高耸的围墙给人们造成了视觉障碍，也减少了城市活力，不仅在一定程度上破坏了日新月异的城市结构，还阻碍了步行街道的发展。当地政府决定对原建筑进行翻修，立即引来了广泛的争论，尤其是当地的设计师，例如建筑师，他们认为新的建筑应该对自然环境起到更好的保护作用，即在建筑中寻求更加平衡的方式。最终，建筑师提出了 "开放场址" 这一方案，即拆除大部分的监狱围墙，建立一个开阔的公共空间——善牧者休闲广场，并设置一个集观赏娱乐为一体的水景设施，既可用做大型集会场地，又可成为市民休闲散步的好去处。这一提议备受瞩目，同时也被认为是最明智的决策。

如今，教堂已不再作为宗教用途，而是作为博物馆向市民开放。教堂的内部结构以及墙上的壁画都已恢复原貌，露台经翻修后被作为户外咖啡馆及露天平台供人们使用，这座旧时的宗教建筑现已成为科尔多瓦市历史文化中不可或缺的一部分。

设计师使善牧者休闲广场朝向城区的两条主要街道和一条林阴大道，使得本来分隔开的地区看上去浑然一体，并在交叉路口处修建了一座喷泉，广场后面是城市的另一个标志性建筑——圣方济各会的教堂。这一景点的开放揭开了宗教建筑的神秘面纱，使其充分展现在市民面前，让市民能够领略到更加令人难忘的城市景观。

主喷泉呈三角形——其中一侧与大楼正面重合；一侧几乎与一座新建的建筑垂直，并与门口的斜坡相邻；另一侧面向大街，呈不规则轮廓，似乎向开阔的城市景观延伸。喷泉周围的铺装从人行道方向至喷泉缓缓向上倾斜，再向下倾斜延至水面，形成尖拱形，充分体现出了科尔多瓦省大坝的建筑风格，同时也形成了天然屏障，使孩子们与水面保持一定距离。这些 "硬质的斜面设计" 与线型混凝土地面交替从水下伸展至人行道上。广场上还设有水泥长凳，人们可以坐在上面聊天、放松心情或观赏喷泉表演；斜面和长凳平行而建，产生一种鲜明的节奏感。音乐喷泉表演开始时，

广场上挤满了孩子和情侣，还有很多人全家一起来观赏；善牧者休闲广场还常常引得三三两两的学生在课间来此漫步，他们有的悠闲地坐着，有的安静地读书，还有的在长凳或郁郁葱葱的草地上享受日光浴。

此外，水景还包括一个小型喷泉，坐落在综合大楼第一层和第二层之间的平台上；喷泉的中心竖起一座雕塑，宛如舞台的布景一样。实际上，这个较小的喷泉是主喷泉的水源，通过瀑布向主喷泉供水。瀑布首先流经首层，在那里形成薄薄的水层，水层向前流淌形成一条小路，上面布满了不规则形状的鹅卵石，人们可以从上面穿过，鹅卵石小路通向五个方向，分别象征着科尔多瓦省的五条河流。

在临近另一条大街、与教堂正面和新建筑相接的开放空间是一片宛如建筑布景的绿坡。沿人行道两侧种植着棕榈树，还有阿根廷著名的城市景观设计师 Diana Cabeza 设计的长凳，很多年轻人常常来此休憩、聊天。一簇簇的棕榈树与新建大楼的混凝土圆柱遥相呼应；无论是棕榈树还是混凝土圆柱，都与圣方济各会的教堂一起展现了该地区哥特式的建筑风格。

善牧者休闲广场呈 L 形环绕在建筑周围，不仅为市民提供了一个自由、休闲、观赏水景、享受日光浴的公共空间，还为政府的城市景观建设增加了有意义的元素。

1　人们在此休憩、观看喷泉水景
2　喷泉有规律地喷发
3　水景是开放空间的主要景观元素
4　场址内新设计了步行道和通道

Paseo del Buen Pastor or Good Shepherd Mall is a recently opened commercial-cultural complex with brand new public spaces in the capital city of Cordoba province, in Argentina. This province is located in the exact geographical centre of the country and it's crossed by five different rivers which compose an important local hydrological system. Taken as a cultural reference to the local landscape, this system was conceptually incorporated to the layout of the open spaces of this development.

The project, completed in 2007, was quite controversial for it renovated the site of a former women prison and a church dating from the beginning of the last century. The church belonged to the Buen Pastor (Good Shepherd) Nuns congregation, built in the typical Spanish colonial style with open loggias and monastic patios; the prison, built adjacent to the church, had a very impersonal architecture, typical of these kinds of buildings with solid tall walls

made with the only objective of denying any contact with the surroundings.

In this case, the walls constituted an especially strong physical and visual obstruction for the urban dynamism of the area and for people because two of them met at one of the most popular corners of this part of the city. Not only the city structure, which had been changing and evolving over time, was in some way partitioned but also the possibility of freely walking and going around was inhibited. When the local government decided to propose this place for rehabilitation it started a big controversial process, especially among local architects and designers: those who stood for a more profound conservationist approach and those who had a more flexible perspective

that sought for a more balanced commission. An initial idea, related to the first approach, was presented and supported by a group of architects who are part of ICOMOS (International Council of Monuments and Sites) but the proposal didn't prosper mainly because it preserved a great deal of the walls. From the very beginning, the idea had been to get rid of those solid walls in some responsible way.

Finally, the architects of the local Department of Architecture proposed the idea of opening up the site by demolishing most parts of the prison and the creation of a void: a plaza with a series of public spaces that would act as a large gathering space and would incorporate a visual recreational water system. This would generate a

completely new space and, in addition, was a considerably less expensive proposal.

The church was secularized and its structure and interior original murals were restored; now opened as a museum, this formerly religious building has become part of the cultural offerings of the site. The monastic patios, on the other hand, were renovated as part of the commercial spaces as outdoor cafes and terraces. The new architecture that provides the spaces for the rest of the cultural rooms and commercial areas is based on the incorporation of pure forms and high-tech aesthetic that generate a big contrast with the existing preserved construction. This, again, was the object of serious opposition and critics.

However, what actually rises as a noticeable wise

1 木柱与棕榈树在视觉上相互呼应

2 绿地与周围的硬质铺装相辅相成

decision—and therefore a successful element of the project—is the one that responded to the idea of generating a large plaza and open spaces for public use. These spaces, which now occupy what used to be the area of the former building of the prison, face two important streets and an important avenue of the quarter. This leaves a series of physically diverse areas, yet visually connected.

The main space, facing the intersection of one of those two streets and the avenue, is dominated by a fountain that runs the full extent of one of the main facades of the complex (the arches of the colonial church's restored loggia) and which is framed from behind by another urban icon of this part of the city, an eye-catching Franciscan cathedral. The opening up of this spot created direct vistas to the formerly hidden religious building, allowing citizens to now have a whole and actually rather dramatic urban perspective.

The main fountain has a somewhat triangular shape with two of its sides leaning against the complex. One coincides with the formerly mentioned church's façade and the second, almost perpendicular, with a new linear box-like construction and an accessing ramp adjacent to it; the third side, which faces the street, has an irregular contour that seems to expand out towards the open urban spot created here. Establishing hard surfaces that softly slope up from the sidewalk level to then slope down towards the water, the layout of this side wants to evoke the typical dams built in this province and, at the same time, generates a physical boundary that keeps kids out of the aquatic surface. These "hard slopes" alternate with a series of linear concrete pieces that begin into the water and extend outwards; when reaching the sidewalk, these pieces act as benches where people sit around, chat and relax or watch the dancing waters show that is set every other hour. The alternation of slopes and linear benches, which are accurately built parallel to each other, produces a fairly evident rhythm and order. When the show is on, the place is usually packed with kids, couples, families and especially groups of students; because the local university has many faculties within this area of the city, these new public spaces are taken over by large groups of them who, before or after classes, sit around, read or sunbathe on the benches near

the water or on the adjoining green slopes.

The aquatic system is completed with a smaller second fountain, located in an intermediary plane between the first and second levels of the complex and therefore disconnected from the pedestrian circuit. As a result, this water feature is perceived as a stage-like space whose centre is occupied by one of the few sculptures that were incorporated into the site. This smaller fountain actually feeds the main one through a waterfall that first reaches the complex's main level, where it shapes a thin layer of flowing water. The layer of water is crossed by a connecting piece that, in this case, recomposes the pathways system by allowing the passage of people from one spot to the next. Five different linear cuts, irregularly placed in this piece, refer to the five rivers of the province

of Cordoba and visually define a series of stepping stones.

When getting closer to the other street, next to the secondary church's façade and part of the new constructions, the open spaces turn into some green slopes that act as a framing podium to the architecture. Planted with some still young native trees this place is also usually dominated by young people gathered in groups. Arranged along the sidewalk onto which these slopes open, appear a row of palms and some benches designed by Diana Cabeza, one of the top urban-furniture designers of Argentina. The palms make a visual and physical connection between this and another paved space which surrounds the slopes and then turns into a semi-roofed area; planted in clusters here, they want to establish a dialogue with a number of concrete circular columns that bear part of the new buildings. All

these vertical elements also visually relate with the lines of the Franciscan cathedral that, at this spot, entirely exhibits its main gothic façade.

All in all, these new open areas are outlined to embrace the architecture. With an overall layout shaped as an "L" whose arms are the ones that embrace the existing restored and new constructions, the public spaces become the one factor that unifies the project. The use and appropriation that locals and visitors have generated in this urban spot adds a very meaningful element to this urban offer: a positive and confirmative response to the necessity of openness and flexibility that this and every city have and, above all, to the necessity of spaces which provide the possibility of movement, freedom, leisure, encounter, sunlight, water and greenery.

1 倾斜的地面一直到达水边

2 喷泉喷发出不同高度的水流，营造出有趣的景象

3 犹如科尔多瓦具有象征意义的水坝

铸造广场

Foundry Square

撰文：SWA Group　　图片提供：Tom Fox　　翻译：丁岩

1　第二建筑区的屋顶花园（以第四建筑区大楼为背景）
2　转角处的郁金香、宽阔的台阶和低矮的坐墙
3　一期完成的两个建筑区

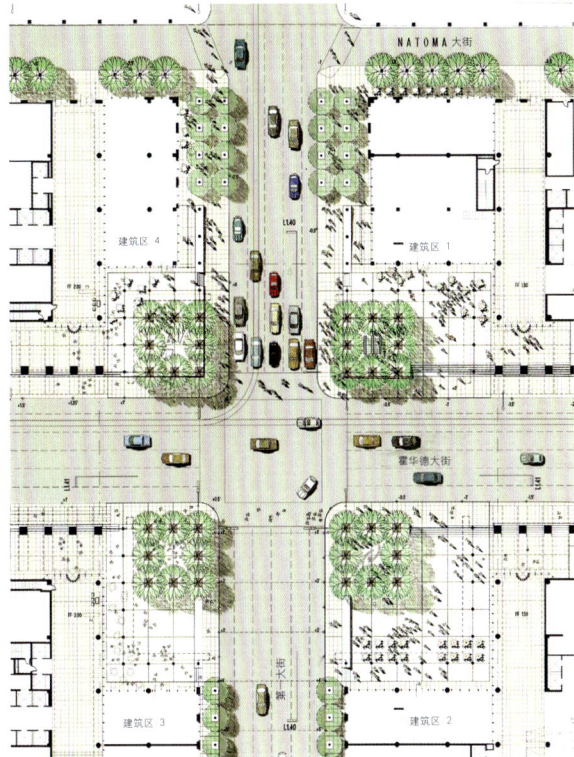

总平面图

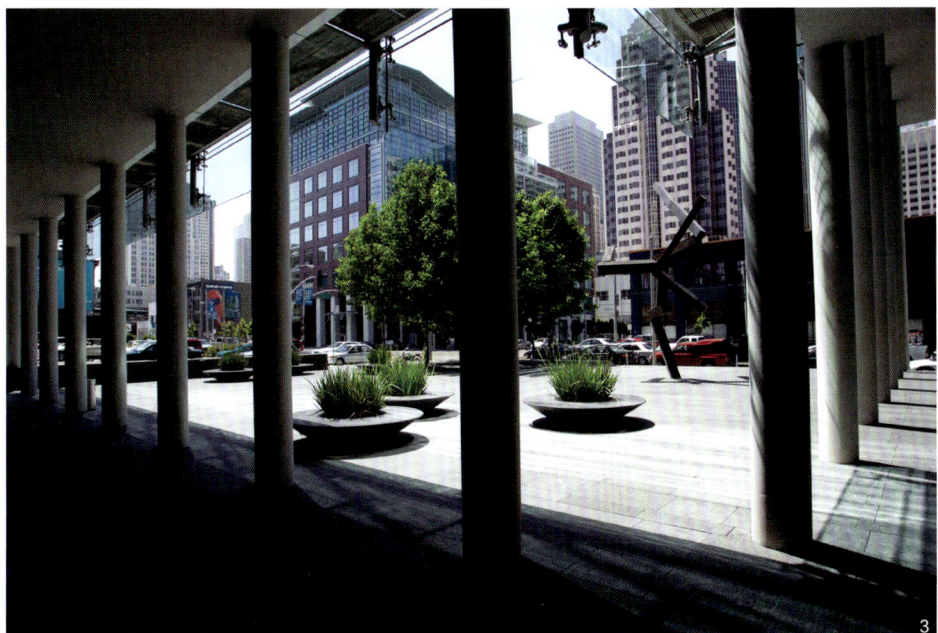

　　铸造广场占地 111 480 ㎡，坐落于旧金山南市场区两条主街——第一大街和霍华德大街的交会处，周围为办公和零售楼群。该项目旨在提供一个与市中心办公楼群相比，面积更大、地板样式更灵活的一流的工作环境，并通过地下通风设施来改善楼群底层的环境。广场的四块建筑区域为公共用地，为营造一处新的城市空间提供了不可多得的机会，也是该项目的关键。

　　SWA 在研究了周围的城市环境后构想了多种设计方案，客户最终选择了能与周围环境相融合的设计方案——街区模式，即四周的建筑作为临街的墙。同时，街区的转角处又各自形成一个广场，人们可以在广场漫步，机动车也可以进入其中。由于该项目与港湾大桥的入口相连，铸造广场自然也成为了上桥的一个入口。客户的目标是为这处办公区营造一个活跃的、以人为本的场所。建筑的外围和广场周围是一些建筑入口的门厅、餐厅和室外咖啡店。

在既定的街道范围内，广场主要是由垂直和水平的结构界定。高 60.96m 的双层玻璃幕墙构成了广场的框架，这种一体的建筑墙体既调节了气候，也增加了办公环境的通透性，成为广场内外空间的过渡区域。在水平层面上，花岗岩铺设的广场成为周围四栋建筑的底座，并且比交叉口处街道的地面高 45.72cm，从而巧妙地将人车分离。由于霍华德大街向东倾斜，宽阔的台阶也呈现同样的坡度，但是底座的高度保持不变。

SWA 提供了城市开放空间设计的研究报告，并且主要负责开放空间的设计和开发。设计的重点是在营造这座公共广场的视觉效果的过程中，使施工与建筑相协调。

原有场地包含几个老的工业建筑，其中一个已经决定予以保留。该建筑在一期工程中进行了改造：地下是酒吧，一层是餐厅，上部是办公区。根据旧金山市停车管理条例，广场与市中心商业区和滨水区一样仅提供有限的停车位，但是为各种交通工具的出入提供了便利的通道。

Foundry Square is a 1.2 million square foot office and retail development that is located at the crossing of two major streets, First Street and Howard Street in San Francisco's South of Market district (SOMA). The goal was to provide a class A work environment featuring larger, more flexible floor plates than downtown office towers, raised floor systems with under-floor ventilation. All four corner parcels were commonly owned, providing a remarkable opportunity for creating a new urban space as the centerpiece for the development.

SWA initially studied the surrounding urban context and formulated several design alterna-tives. After review, the client approved the design approach that juxtaposes the existing context, a block pattern in which buildings form the street wall and hold the corners. In contrast, Foundry Square's open corners collectively form a square, which embraces both the pedestrian movements along with the perimeters

and the street traffic in the middle. Being adjacent to the Bay Bridge access, Foundry Square creates a "threshold" that one passes through in moving onto the bridge. The client's goal was to create the active people-place for the new multimedia based work place. In conjunction with the architectural enclosure, the base of the square is to be lined with building lobbies, restaurants, and outdoor cafes.

Given the street scale, the square is essentially defined by its vertical and horizontal components. The building's 200-foot dual-glaze tech walls frame the square. This unifying building wall element offers climate control and transparency to the office environments and establishes an arcade at the transition between interior and exterior spaces. At the horizontal plane, granite plazas span the intersection to establish a taut plinth for all four buildings. The plinth is 18 inches above the street at the intersection to create a subtle separation between people and cars. As Howard Street slopes

down to the east, broad steps take up the grade change but the plinth elevation, set level on pedestals over structure, remains constant.

SWA was an integral part of the interactive design team process. SWA provided urban design studies for the open space and was primarily responsible for the design and development of the open space. The focus was to seamlessly integrate the site work with architecture in realizing the vision of the Public Square.

The existing site contained several old industrial buildings, one of which was determined to be worthy of preservation. As a result it has been integrated into the Phase 1 building where it will serve as a below-grade bar, at-grade restaurant, and office spaces above. Per a City of San Francisco parking ordinance, the Square has limited parking provided but convenient access to multiple modes of transit as well as the central business district and the waterfront.

西蒙·弗雷泽大学的新衣 —— Town+Gown 社区广场

New Dress for Simon Fraser University — Town + Gown Community Square

撰文：Ann Jackson　　　图片提供：PWL 景观设计公司　　　翻译：牟誉

概念透视图

该项目所在的社区是加拿大西蒙·弗雷泽大学内典型的可持续型社区,为上万名居民提供了一片充满活力的公共空间。这个用途广泛的大学社区坐落于不列颠哥伦比亚省伯纳比市伯纳比山山顶,集居住、购物、服务和生活设施等于一体。该项目在提供多样的社会和商业机会的同时,坚持生态建筑、保护自然居住环境、选择可持续性交通方式并完善社区设施。为此,Town+Gown 社区广场设在了现有的大学设施、公共交通站和新开发区域的商业中心的交叉口处。

该项目加强了校园与新社区的自然及人文联系。这种"纽带"作用是设计的核心理念,并已被证明是该项目成功的一个关键因素。因此,广场位于社区中央,与市区商业区及周边 300 万平方米的森林景观相毗邻。PWL 景观设计公司参与了概念设计、细部设计和营造等工作。

该项目根据坡度变化分为两个主要部分——间隔均等的水泥台阶为户外表演、正式活动及聚会提供了更多的坐位;平台、石阶、邻近的建筑物与造型各异的水景、树木、校园建筑和街景构成一系列户外空间,可移动的坐椅、植栽容器和树木也营造出多样的氛围和空间。设计团队还利用坡度变化扩大了邻近建筑物下的停车场,以容纳一个泵房和储物区。

现在,这里已成为节日庆祝、公共表演、学术演讲、观光和聚会的重要场所。设计采用了加固路面和可移动安全柱,这样可以允许机动车进入广场内较低的地段,以便装卸音响设备、备用坐席或戏剧表演的道具;在广场的大型开放空间上,一座桥跨水景而建,使这

1 喷泉夜景
2 从广场西南角看广场中央和左侧水景

规划图

023

里成为既开放又具有私密性的空间；邻近街景的信息亭便于社区居民张贴和分享信息。

广场的规模和比例都参考了艾里克森和梅西建筑师事务所在 1964 年西蒙·弗雷泽大学总体规划中对公共空间的设想。水景墙、挡土墙和构筑物等细节提升了校园原有的建筑环境。整体而言，广场的细部设计整洁而优雅，设计师通过复杂的路面铺装图案突显了广场的重要性，吸引行人驻足于校园的这片中心地带。植物烘托出自然而繁茂的环境效果，与硬质路面形成对比。占有重要地位的水景则充分利用了场地的坡度变化，俨然山体的一部分，成为广场西侧的视觉标志。水景墙结合本土森林植被，虚拟了加拿大西海岸的地形和温带雨林景观，成为校园数十年来的一个重要的代表性元素。以本土森林植被为背景的阶梯形瀑布构成了令人难忘的新城市空间的奇妙背景。

该项目从三个方面遵循可持续发展的原则——社会、经济和环境，而首要目的是达到社会领域的和谐统一。设计师也从若干领域考虑环境问题，如将雨水管理与周围的街景设计结合在一起，包括渗透性较好的铺装路面和当地一些质地坚固的材料等。另外，对人们而言保留山顶上的夜空和星光也尤为重要，为此设计师精心调整了照明设计，营造出一个别具一格且安全的公共空间。出于对环境保护和维护的考虑，喷泉中的水也经过了加氧处理；广场周围的植物都是从周边地区移植过来的，包括乔木、灌木和地表植被等。这样一来，一片小型森林便被引入社区的中心地带，呈现在大学校园的入口处。

该项目的设计融合了现代主义的建筑设计理念和社区可持续发展的要求，是温哥华地区中少数几个能实现这一要求的项目之一。Town+Gown 社区广场作为新大学城社区里具有代表性的设计，成为了这个可持续型社区的鲜明特征，得到了居民、学生及设计界人士的广泛认同。

东西向剖面图

南北向剖面图

1 广场一角

2 在清晨的阳光中享受静谧时刻的学生

3 池底材料近景

Town + Gown Square provides a vibrant and important public space for UniverCity, a model sustainable community of 10,000 people at Simon Fraser University. Located at the top of Burnaby Mountain in Burnaby, British Columbia, this mixed-use community provides a range of housing choices, shops, services, and amenities for the university community. The primary objective for this community is to provide diverse social and economic opportunities while maintaining green building and habitat preservation requirements, sustainable transportation options, and community amenities.

To this end, Town + Gown Square was located at the intersection of the existing university facilities, the public transportation hub, and the commercial centre of the new UniverCity development.

The goal of the square was to provide a physical and social connection between the campus and new residential neighbourhoods. This idea of bridging is a central theme and one that has proved to be a key ingredient in the success of the project. This central location also juxtaposes the more urban High Street and the surrounding 300 hectare

3

forest landscape. PWL was involved in the project from the beginning of the conceptual design phase through detailed design and construction of the square.

The Square is divided into two main terraces that are separated by grade changes. The concrete steps feature wide blocks at regular intervals that provide extra seating for outdoor performances, events, or informal gatherings. The terraces, steps, and adjacent building create a collection of outdoor rooms with different views of the water feature, surrounding forest, university buildings, and adjacent streetscape. Moveable seating, planters, and trees also provide a range of climates and spaces for different uses and

seasons. The design team also took advantage of the grade change to extend the parking garage of the adjacent building under the Square to accommodate both a pump house and storage area.

The Square is now the venue for festivals, public performances, academic discourse, tours, and informal gatherings. Reinforced paving and removable bollards were incorporated into the design to accommodate vehicle entry into the lower portion of the square for the unloading of sound equipment, additional seating, or theatre apparatus. The large open space of the square functions as an outdoor amphitheatre with a bridge over a portion of the

water feature that also doubles as a covered performance space. An information kiosk adjacent to the streetscape provides a space for announcements and events, and allows community members to post and share information.

Both the scale and proportion of the square reference the public spaces originally envisioned by Erikson/Massey Architects in their 1964 Master Plan for Simon Fraser University. The detailing of the water feature walls, stone retaining walls, and structures reinforces the low horizontal expression of the existing campus architecture. Overall, the detailing of the square is clean and elegant, which allowed for a more complicated paving pattern that announces the

square's importance and draws pedestrians into this central campus space. The plants were chosen for a natural and lush appearance in order to provide contrast to the hard surfaces. The dominant water feature takes advantage of the site's changing grades and visually anchors the west side of the square, appearing to be part of the mountainside. Combined with the native forest planting behind, the water feature celebrates the Canadian west coast topography and temperate rainforest that have been an important and defining element of the campus for decades. The cascading fountain feature with its backdrop of native forest planting provides a dramatic backdrop to this memorable new urban space at the top of Burnaby Mountain.

The development of Town + Gown Square addresses all three areas of sustainability—social, economic, and environmental—and while its primary mandate is in the realm of social cohesion and community, the design addresses environmental issues in a number of areas. PWL integrated rainwater management features within the adjacent streetscape and the design included permeable pavers, as well a range of locally sourced and durable materials. It was important to preserve the darkness and the starlight of the night sky at the top of the mountain, so the lighting design carefully modulates illumination to create an attractive and secure public space. The water in the fountain is treated with ozone to address both maintenance requirements and environmental considerations. The surrounding planting includes trees, shrubs and ground covers transplanted from the surrounding development sites. In this way, the forest is brought into the heart of the new residential community and to the entrance of the university campus.

This is one of a few projects in the Metro Vancouver area that has provided the chance to articulate the relationship between a significant modernist architectural design and a new environmentally sustainable residential community. As the main feature in the new UniverCity neighbourhood, Town + Gown Square strengthens the identity of this model sustainable community. It has been well received by residents and students, and by the design community at large.

伦敦温布利表演竞技场广场

Arena Square, Wembley, London

撰文：Randle Siddeley   翻译：王玲

温布利重建规划是伦敦的布伦特自治市的重点项目。现已竣工的新温布利体育场将各种世界级的赛事重新带回到温布利。2002 年末，布伦特自治市发起"我心目中的新温布利"活动，人们对该项目寄予厚望，希望营造布伦特的社区中心、弱化公共空间和私人空间之间的界限，打造一个人们引以为傲、热情向往的场所。表演竞技场的改造耗资 3500 万英镑，它与体育场之间的区域是温布利首个设有互动喷泉和灯光效果的公共区域。

表演竞技场广场改造是新温布利重建规划的第一阶段，占地面积约为 15 000 平方米。表演竞技场原来的杂物院被改造成一个与 20 世纪 30 年代硬质景观园林相呼应的阶梯式广场，硬质景观园林最初是 1924 年帝国展览会的一部分。表演竞技场广场通过举办各种季节性文化活动、热身表演和庆祝活动来满足人们的不同需求。重建规划还涵盖了另外 270 000 ㎡的占地，包括一个节水缸、特色设计、照明、道路和西班牙式阶梯。

表演竞技场的建筑和通向体育场的交通动线都被重新设计，营造出一个全新的文化社交中心。根据在 20 世纪 20 年代做的最初设计，表演竞技场与一个新

建的世界级高品质的广场相通，该广场不仅在体育场和表演竞技场之间提供了活动场所，而且承载着表演竞技场的各种户外活动。表演竞技场广场是通往温布利公园大道的入口，公园大道在未来的开发中将连接温布利高速路和温布利公园车站。设计的主要原则：

· 创建流畅、便捷的主要公共空间和高品质的交通动线；

· 创建强烈的场地特征和场所感；

· 创建一处通行便捷、管理良好、备受欢迎的公共空间；

· 按照《体育场安全绿色指南》连接温布利体育场公共大厅和表演竞技场；

· 为新的国家体育场营造出高品质的周边环境；

· 优化已列入遗产保护（Ⅱ级保护）的温布利体育场；

· 充分利用该地区便利的公共交通可达性，全面释放重建潜力；

· 展现该地区白天、夜晚和周末的生机与活力；

· 促进与新的国家体育场、开发区域以及温布利的整体经济联系；

· 创造一个高品质的艺术、文化和娱乐中心；

· 为西部总体开发提供节水储备设施。

项目设计的动力源于将表演竞技场广场建成一个世界级的伦敦广场，使之可以与沉稳优雅的、欧洲最好的特拉法加广场相媲美。广场上的活动可以灵活转换，特别是在大型活动前后，仿佛一按按钮，歌舞等户外活动由地而生，吸引游客驻足此处。

设计师力图将表演竞技场广场打造成一个全新的文化胜地——"北伦敦的南岸"。设计的关键是确保场地的灵活性，使之能够随着环境的发展满足不同的功能需求。

设计不仅需要在短期内完成，还要适应地块今后10年的逐步发展，兼顾到未来的总体规划。该项目是一个在商业还未充分开发，但公共空间已被设计和建造并取得良好效果的典范。RSA 将继续负责公共空间的总体规划。

The regeneration of Wembley is central to the London Borough of Brent's aspirations for the Borough. New Wembley Stadium is now complete and bringing world-class events back to Wembley. The LBB produced "Our Vision for a New Wembley" in late 2002, and the aspirations particularly relevant to Arena Square are: providing a community focus for Brent, blurring the boundaries of public and private space and creating a place where people are proud to live and eager to visit. The Arena has had a £35 million transformation. The area between the Stadium and the arena is Wembley's first public space that has interactive fountains and light shows.

Arena Square is the first completed phase of the New Wembley regeneration scheme, covering an area of approximately 1.5ha. The existing service yard of the Arena has been converted into a stepped amphitheatre that echoes the 1930s' hard landscape garden that originally formed part of the 1924 Empire Exhibition. The square meets the requirements of the client by providing a dramatic setting for seasonal and cultural events, Arena warm-up shows and celebrations. The masterplan area covers a further 27

1　广场上的水景
　　（图片提供：Randle Siddeley Associates）
2、3　广场全新夜景
　　（图片提供：Randle Siddeley Associates）
4　水景
　　（图片提供：Randle Siddeley Associates）

hectares including an attenuation tank, special features, lighting, roads and Spanish steps.

The Arena building was refurbished and re-orientated towards the Stadium, creating a hub for cultural and community activity. In the spirit of the 1920's original design, the Arena would open onto a new square, which would be of a "world class" quality and provide a powerful new focus for activity between the Stadium and Arena, as well as accommodating the outdoor events planned for Arena Square. The Square was to serve as the gateway to Wembley Park Boulevard which will through the future development linking Wembley High Road, and the Wembley Park Station. The key principles for the design were:

• To create a character of outstanding sequences of major public spaces and high quality routes.

• To create a strong identity and sense of place.

• To achieve a public space of wide appeal, which is permeable, legible and well managed.

• To provide a link between the Wembley stadium public concourse level and arena designed in accordance

with " Stadium safety Green Guide".

• To deliver a first class setting for the new National Stadium.

• To enhance the setting of the listed Wembley Arena (Grade 2 listed).

• To make the best use of the high public transport accessibility levels of the area in maximising regeneration potential.

• To generate vibrancy and vitality in the day, in the evening and at the weekend.

• To stimulate economic linkages with the new National Stadium, within the development area and with Wembley as a whole.

• To create an enhanced focus for arts, culture and entertainment.

• To provide a water attenuation storage facility for the western masterplan development.

The inspiration was the creation of a world-class London square to rival the scale of the Trafalgar Square in the spirit of the calm, sophisticated elegance of the best European squares; with the flexibility to be transformed at the push of a button (literally) to an all singing and dancing outdoor venue to increase dwell time of the visiting public, particularly before and after events.

Arena Square is intended as a new cultural destination, a "South Bank at North London". Central to the vision for Arena Square was to ensure that the design had flexibility, thereby allowing for a range of different functions within the square as the surrounds developed.

The design had to work within the short term whilst also accommodating the future master plan design as the parcels of land are gradually developed over the next 10 years. This is an example of the positive effects of public realm being designed and built in advance of a commercial development. RSA have a continuing appointment as masterplanners responsible for the public realm.

福尔特文格勒花园再设计

Furtwangler Garden Redesigned

撰文：Auboeck+Karasz　　图片提供：Andrew Phelps　　翻译：王玲

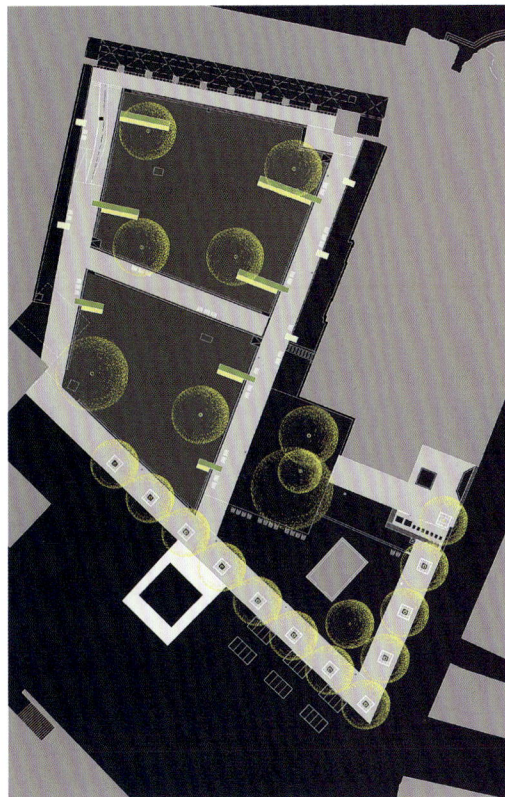

1 夜景
2 步道
3 萨尔茨堡
4 从艺术节礼堂方向望去的景观

规划图

从中世纪开始，萨尔茨堡市便将街道、广场、庭院和通道等公共空间融为一体，形成了独特的城市景观。福尔特文格勒花园在重新规划时加入了一种全新的趣味元素——城市中心的现代花园式广场。

该花园坐落于学院教堂、古老的大学和萨尔茨堡艺术节礼堂之间，展现了对 17 世纪和 18 世纪萨尔茨堡市由墙体围合而成的花园的现代诠释：由淡绿色的石英岩和白色的沙砾碎石铺成的宽阔散步道宛如两块低洼草地的边框。

广场周边的绿化带是一排椴树丛，为这处公共空间勾画出全新的轮廓；阡陌纵横的散步道也是很好的休息场所。

错落有致的树篱将种着槐树的草坪和散步道连接起来。游客坐在花园中的椅子上休息时的感觉丝毫不逊色于在露天平台上的感觉，他们在城市的中心观赏着绿地，享受着宁静。弗里茨·沃特鲁巴、埃米利奥·葛雷高、杰阿柯莫·曼祖、安塞姆·基弗的艺术作品分布在花园中，使整体效果更加完美。这里还可以作为萨尔茨堡艺术节礼堂的门厅。

Since medieaval times the public spaces of Salzburg merge from streets, squares, courtyards and passages to a unique urban setting. The redesign of the Furtwängler garden adds an interesting new element: a contemporary garden-square in the city centre.

Situated between the Kollegienkirche, the old university and the Salzburg Festival Hall the garden unfolds as a modern interpretation of Salzburgs inner gardens of the 17th and 18th century bordered by walls: A wide framelike path—partly in light green quartzite, partly in white grave—constitutes the enclosure for two sunken lawn fields.

A broad of limetrees form an edge towards the bordering square creating a new outline in the urban fabric. This net of promenades serves as a relax area as well.

The deepened lawns planted with sophora trees are linked to the boarding promenades by scenographically shaped hedges. The visitors are relaxing on garden chairs comparable to a veranda deck. They are looking towards the green fields, enjoying peaceful silence in the heart of the city. Art works of Fritz Wotruba, Emilio Greco, Giacomo Manzù and Anselm Kiefer complete the ensemble, which serves equally as a foyer for the Salzburg festival.

1、2 艺术节礼堂
3 大学礼堂

音乐中心表演广场

Music Center Performance Square

撰文：Peter McGuckin　　　图片提供：SLR Consulting Ltd.　　　翻译：刘建明

诺曼·福斯特爵士的代表作塞奇·盖茨黑德音乐中心位于泰恩桥和波罗的海当代艺术中心的正中央。音乐中心俯瞰着盖茨黑德千年桥，在以艺术为主导思想的重建项目群中，这座桥是第三个子项目，也正是这座桥令盖茨黑德在过去的十年中冠绝英伦。

盖茨黑德码头二期是一个名为GQ2的综合开发项目，集零售、休闲和公寓为一体。这个新兴的开发项目与塞奇·盖茨黑德之间的空间便是极具名人效应的表演广场。

该项目在总体规划中设计了一系列独具特色且功能各异的空间，这些空间从整体上建立了从盖茨黑德市政中心到"少女之路"以及与盖茨黑德码头之间的联系。

通达性与渗透性是盖茨黑德码头设计成功的关键。

尽管塞奇·盖茨黑德所处斜坡的侧翼与泰恩河河岸之间存在着高差，但经过精心设计后，人们可从西面和盖茨黑德市政中心轻松进入会场。这一系列空间的核心区域在塞奇·盖茨黑德的东侧，那里是特别保留的活动空间。此区域具有极大的开发潜力，它将成为观赏河岸景观以及纽卡斯尔市天际线最理想的公共空间。

设计团队所面临的挑战是要设计一处极具想像力的开放空间，使之具有一定的灵活性，能够避免对公开活动、音乐会等造成空间束缚；同时可供游人休憩，作为GQ2的开发计划中不可多得的一处休闲空间。

这一享誉颇高的设计旨在以一种激动人心的现代方式来诠释渗透性、通达性和灵活性。虽然表演广场无需与塞奇·盖茨黑德音乐中心争奇斗艳，但在设计上也需要占据足够强势的地位，以防止对面正在规划

的酒店通过蚕食广场边缘的方式来弱化广场空间。

除此之外，设计团队还设定了另一目标——该项目必须有一个独一无二的主题，即一个与邻近的塞奇·盖茨黑德音乐中心相关的主题。

设计团队一致认为人居空间应以安全性为主。因此，对表演广场的定位已经远远超出了只是一系列空间（包括"少女步道"）的一个组成部分；"少女步道"是南向连接盖茨黑德市中心与波罗的海产业园区，北向连接泰恩河的一条重要的步道。塞奇·盖茨黑德音乐中心的巨型玻璃幕墙构成了令人瞩目的西向背景幕；广场东侧地势高差较大，此处的临时台阶将被新的GQ2开发项目所取代。

表演广场主要的设计灵感来自于塞奇·盖茨黑德音乐中心的主要功能。象征声波的图案从中央一点向

四周发散，而障碍物所形成的干扰使整个项目看起来更加生动形象。

通过运用各种不同材料可以更好地诠释"声波"的设计。从南面至广场正面是舒缓的植草斜坡，而圆形阶梯剧场的较高位置则是陡峭的草坪阶梯。这些阶梯依照声波的形状设计，逐渐向剧场的中心靠近，较低的中心位置被设计成花岗岩阶梯。

圆形剧场的正前方是一个开放的广场，设计师选用粉色的花岗岩和银灰色的混凝土铺装材料来修饰圆形的涟漪状波纹。这样的设计可以充分引起共鸣，"波纹"覆盖整个圆形剧场，在阶梯较低的位置处纵横交错的镶边铺装形成"声波干扰屏障"。

表演广场的每个层面都可以用来演出和集会，例如可以供大型旅游团体在此观赏塞奇·盖茨黑德音乐中心以及泰恩河北岸壮观的景色，另外还可以作为大型活动的场地。

对声波和干扰图案的研究早在设计之前就已经展开。虽然这种情况是逻辑主导了美学，但在表演广场多次成功举行了正式活动和户外广播节目后，人们惊喜地发现表演广场不仅在视觉上美观和谐，同时也实现了空间声学理论在现实中的应用。

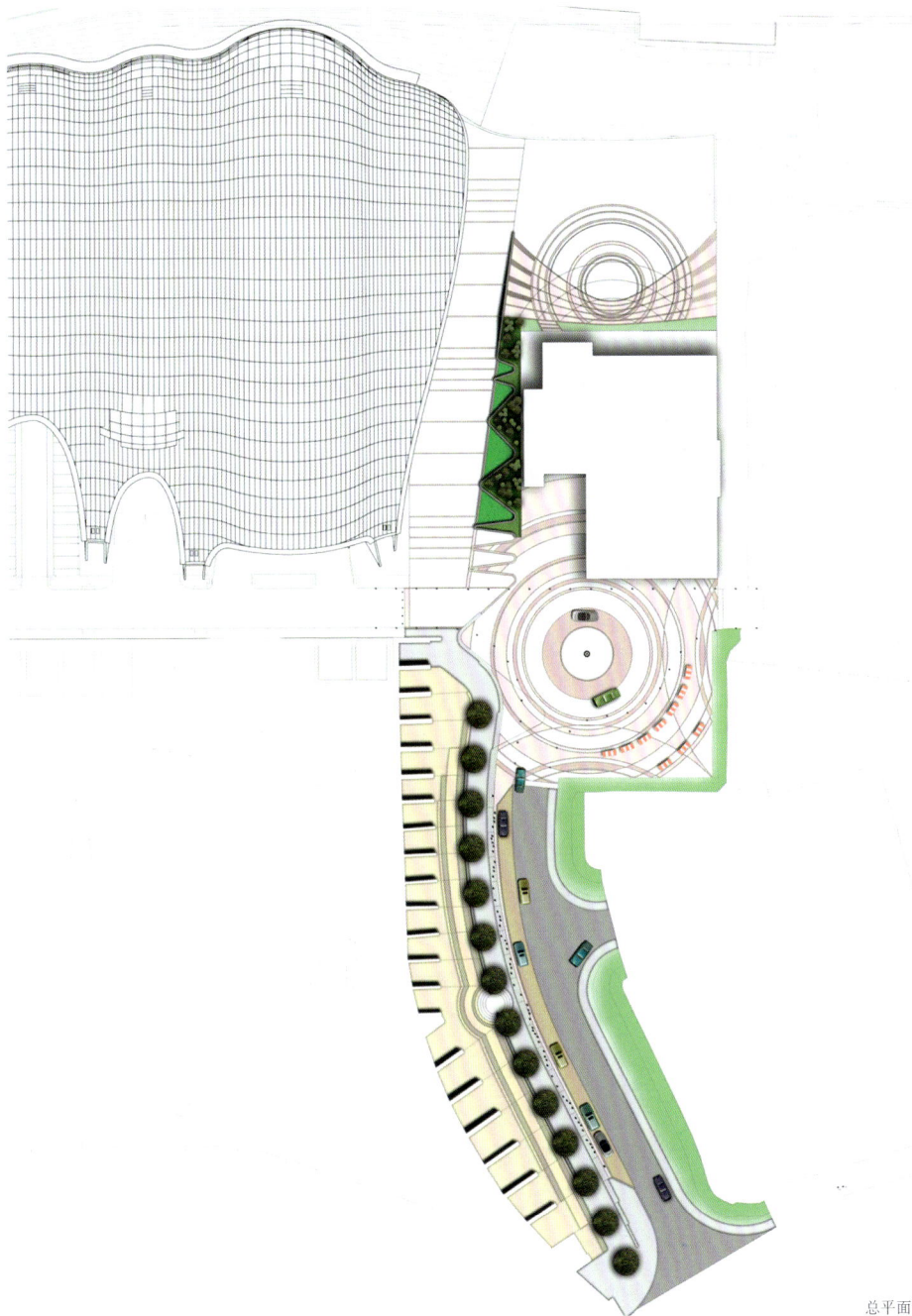

1 表演广场 1

总平面图

Sir Norman Foster's striking Sage Gateshead music centre lies midway between the Tyne Bridge and the BALTIC Centre for Contemporary Art, overlooking the Gateshead Millennium Bridge, the third member of an impressive clutch of arts led regeneration projects which Gateshead has championed in the last decade.

Gateshead Quays phase 2 is a mix of retail, leisure, and apartments now known as GQ2. The space between this forthcoming development and The Sage Gateshead is Performance Square reflecting the purpose of its prestige neighbour.

The project required the development of a masterplan to provide a series of linked spaces of varying character and function. These spaces would collectively form a connection from Gateshead town centre, through Maiden's Walk to Gateshead Quays.

Accessibility and permeability are critical to the success of Gateshead Quays. Despite challenging level changes between the shoulder of the slope (which The Sage Gateshead sits on) and the banks of the Tyne, the venue is readily accessible from the west and Gateshead town centre. A principal area in this series of linked spaces is the east side of The Sage Gateshead which was reserved as an event space. This area held great potential to be developed as a key piece of public realm with magnificent views over the river and Newcastle's skyline.

The challenge to SLR Consulting was to provide an

1、2　表演竞技场

3　塞奇·盖茨黑德音乐中心

4　通往塞奇·盖茨黑德音乐中心的表演场

imaginative open space, one which was sufficiently flexible to avoid constraining open air events, concerts etc. and significantly a space which would sit comfortably alongside an as yet unknown GQ2 development footprint.

This prestigious project aimed to interpret permeability, accessibility and flexibility in an exciting contemporary fashion. Without vying with The Sage Gateshead, Performance Square would nevertheless need to assert itself sufficiently strongly to deter a planned hotel on the opposite side of the space from nibbling away at the edges.

In addition to this, SLR also set their own requirement; the space must have a distinctive theme—one which would relate to the performances in the neighbouring Sage Gateshead music centre.

The team felt strongly that a peopled space would be a safer space. Consequently, Performance Square is but one component in a series of linked spaces including Maiden's Walk an important pedestrian route leading south to Gateshead Town Centre and to Baltic Business Park, and north down to the river Tyne. The Sage Gateshead's large curtain glass facade is an impressive backdrop to the west. The land presently falls steeply away on the east side where a flight of temporary steps will be replaced by the new GQ2 development.

The Sage Gateshead's prime function as a music centre became the main inspiration for Performance Square. The notion of sound waves emanating from a central point, and the effect which obstacles might have in interfering with the pattern of these was considered a fitting metaphor.

Sound waves also lend themselves well to being interpreted in a variety of materials. A grassy lee slope rises gently from the south to cascade down a steep scarp slope in a series of lawned steps which mark the upper tiers of an amphitheatre. These steps are cut to the form of sound waves which become tighter toward the focus of the theatre where they give way to a lower tier of granite steps.

A broad plaza opens up in front of the amphitheatre. Here bands of pink granite and silver grey concrete paving are picked out in a circular rippling wave which is sufficiently resonant to dominate the amphitheatre ripple with the "interference" revealed as vertical and horizontal pecks and bands across the lower steps.

Performance Square appears to be performing well on a variety of levels, as a gathering space, a place for large tour groups to assemble and admire The Sage Gateshead and the magnificent view north across the Tyne, and also as a major event space accommodating large crowds.

Research into the behaviour of sound waves and interference patterns was conducted before design began. Although logic led aesthetics on this occasion, a number of events and outside broadcasts have since been held in Performance Square and it is gratifying to find that the acoustics of the space work as well in reality as the space harmonises visually.

朗斯代尔滑板广场

Lonsdale Skate Plaza

撰文 / 图片提供：Jim Barnum Derek DeLand B. Sc B. Arch　　翻译：刘丹春

该项目在许多方面具有创新性：它是第一个试图成为不仅是滑板场的滑板场；它是第一个滑板广场。滑板广场是一处造型美观、建筑可靠的公共场所，为滑板人员提供聚会和练习的场地；也是一处集都市庭院、公园和行人广场于一体的场所，如费城的"爱心公园"、旧金山的"加斯汀·赫曼广场"、巴塞罗那的"桑兹火车站广场"，都是标志性的滑板人的城市空间。

滑板不仅仅是一项体育运动，还具有创造性、进取性和自由精神。但是传统的滑板场空间狭窄、人造色彩浓重，也形成了对滑板精神的反冲，其拘束、枯燥乏味和单调的设计使滑板人觉得他们好像被困在动物园中，因此，传统的滑板场大多满足不了街头滑板人的需要。

滑板广场可以解决以上问题。滑板广场由楼梯、栏杆、花盆盒、长椅以及其他元素构成，它不仅在形式和功能上，还在感觉上复制（或修改，以提高可滑行性）了以上元素。若要将生命与"真实性"赋予广场，使其成为像市区公共广场那样的"真正的"公共空间，绿化空间以及对各种类型材料的分层使用都非常有帮助，如对石头、砖、钢材、混凝土块、花岗岩、酸蚀混凝土等的运用。另外，设计融入了创新的雕刻形式而不是采用简单的楼梯和栏杆。细节设计包括暗纹与暗花的使用，这也增加了设计的复杂程度。

该项目建造于原停车场之上，包括本地植被（一个大范围的生态湿地区域，解决几乎所有设施的地表排水问题）、雕塑、仿效北岸青山的五彩花岗岩波浪的背景、取代混凝土的钢质滑板场表面以及多样化的"滑行平台"或直立墙，细节设计包括原石镶面、侧墙和悬壁结构。当然，其核心部分由上文提到的复杂的都市景观形式（楼梯和栏杆等）构成，提高了逻辑上的"滑板运动流程"和连续性。

广场西部边缘向上提升使之与相邻的人行道形成阶梯状关系。而下沉广场形成了一个露天剧场场景，与木板人行道和其他观赏空间一起，将非滑板公众引入到这里，实现了使广场成为"真正的公共空间"而不仅仅是一个滑板场的目标。

The Lonsdale Skate Plaza was revolutionary: the first skatepark that attempted to not be a skatepark. It was the first Skate Plaza. A Skate Plaza is an aesthetically pleasing, architecturally informed public place for skateboarders to congregate and practice their sport/lifestyle; a place which replicates metropolitan courtyards, parks and pedestrian plazas such as Philadelphia's "Love Park", San Francisco's "Justin Herman Plaza", and Barcelona's "Estacio-Sants" train station plaza, to name but a few urban spaces that are iconic to skateboarders around the world.

Experienced skaters will tell you that skating is more than just a sport; that it has a lifestyle and soul aligned with creativity, progression and freedom. Thus, with the rise of traditional skateparks which are contrived, confined and restrictive came a groundswell backlash against them. The controlled, sterile, monotone character of traditional skateparks made skaters feel like they were trapped in a zoo. The core street skater's needs were not met by traditional skateparks.

The solution to this problem statement is the Skate Plaza. A Plaza is comprised of elements that would be found in the "real streets" including stairs, railings, planter boxes, benches and more. It must replicate (or modify to improve "skateability") these items not only in form and function, but also in feeling. To imbue the Plaza with a sense of life and "realness", like a "real" public space such as a downtown public plaza, the inclusion of planting throughout the space is helpful, as is a layered use of varied material types including stone, brick, steel, split-faced concrete blocks, granite, acid-etched concrete

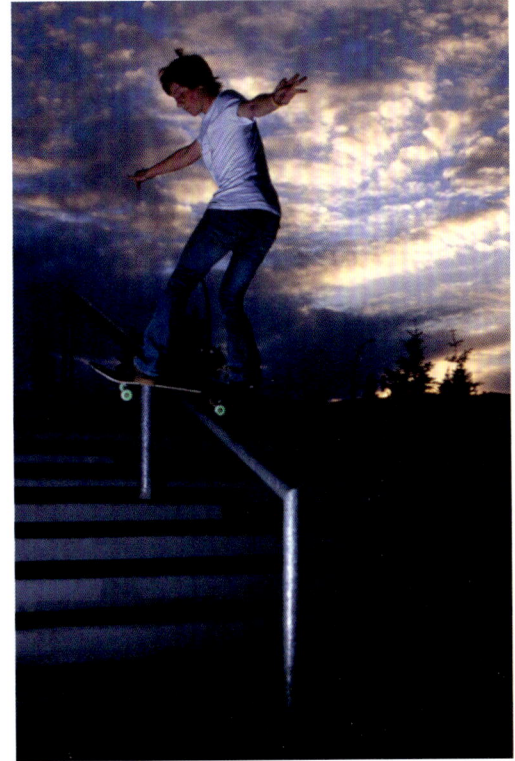

and more. Detailing including the use of shadow lines and reveals adds an additional level of sophistication. Still further, creative sculptural forms beyond simple stairs and railings should be included.

Built in a former parking lot, the Lonsdale Plaza includes significant native planting (including a large bioswale area that handles nearly all of the facility's surface drainage); sculptural, colourful waves which, capped in granite, emulate the backdrop of the Northshore mountains; steel skate surfaces instead of the usual concrete; and multiple "skate ledge" or upstand wall details including native stone facing, reveals and overhangs. Of course, at its core, it is formed by an intricate system of the urban forms mentioned above (stairs, rails etc.) which promotes logical "skateboarding flow" and connectivity.

The elevation of the western edge of the Plaza cascades down in relationship with the neighbouring pedestrian sidewalk. Sinking the plaza into the site creates an amphitheatre setting. Coupled with an elevated boardwalk and other non-threatening viewing spaces, the plaza achieves its goal of being "real public space", not just a skatepark, by inviting the non-skating public into the milieu.

布鲁塞尔弗拉基广场

Place Flagey, Brussels

撰文：Latz+Partner　　图片提供：Latz+Partner　Bernard Capelle　Serge Brison

翻译：王琳

在比利时的首都布鲁塞尔有一个文化氛围十分浓郁的区域——伊克塞尔。设计师在新建的地下车库和地下雨水蓄水池的上方，重新设计了一个城市广场。

场地原来有一个附近水系中面积最大的湖，但19世纪中叶以后，这个湖的部分水域被填埋，在圣克鲁斯（Saint-Croix）教堂前兴建了闻名的圣克鲁斯广场和步道，并成为了当时该地区的中心地带。在20世纪70年代以后，在广场的另一侧兴建了 Foyer Ixellois 综合社区，包括比利时国家电台和一些住宅。之后，地下雨水蓄水池和地下停车场的兴建为该广场的重新设计提供了契机，并借此机会将地面的机动车道和停车处进行整合改造。

考虑到伊克塞尔地区多样化的需求以及轻松愉快的氛围，设计师只有一种选择可以使该地区完美转型，即营造出一个所有人都可以享受的、氛围轻松且和谐的开放空间——增强其功能性与共存性；提供一个"中性"空间，刚柔并济、兼顾长短期使用功能；开启一个能够让人们自由使用的多功能平台；摒弃太过强势的功能性设计语言，从而避免塑造主题式的场地形式。

设计师在设计中遇到了一些挑战——需要整合现存场地上的各个因素，包括已有的环绕广场运行的交通干线、带棚的有轨电车站和公共汽车站、地下车库入口，以及其他一些为行人服务的基础设施。该场所不仅为人们提供游玩、休憩的场所，还可以作为每周一次的水果蔬菜集市。

该项目的设计理念融合了当地独特的文化和经验，同时，市民的参与也贯穿于设计中的每个阶段。当地市民的参与方式多种多样，其中包括征求公众意见、开展各种主题的培训班或研讨会，使他们成为设计师们的"专家顾问团"；而他们的意见被融入到项目的规划设计和开发之中，比如系统的选择、广场活动的设计等。在使用过程中，他们仍积极参与广场的维护工作。

在广场上，水珠四溅的喷泉，三三两两排列的各种树木，灰蓝相间的地面结构，所有设计都在映射着场地历史的同时又与附近公园和其他公共空间（佩索亚广场和圣克鲁斯广场）形成了有机的融合。广场植被的设计体现了自然与人工元素的结合：婀娜的柳树、挺拔的枫树与湖边的植被交相辉映；柳树、栗子树则与北面现有的绿地形成鲜明的对比；电车站与公共汽车站天篷的选材既有质量较轻的玻璃，也有立体结实的钢架。

喷泉被镶嵌在广场的石质表面上。四溅的水花为这个略显乏味的石材铺装增添了些许童趣，在炎炎的夏日里，为游人增添了一丝凉意。

因为广场的石材铺装和附近街道的沥青路面均属同一色系，所以基础设施的设计就要起到区分街道与广场的作用。设计师将"流动的长椅"系列模块安置在广场的西北侧和东南侧，高低不同的椅背像音符一

样排列在广场的两侧，将广场清晰地呈现出来。这些长椅的设计既保证了游客的安全，又使游客可以按照个人的喜好选择朝阳或者背阴的椅子，且背阴的椅子都贴心地安置在树下。值得一提的是，这里的人们都来自五湖四海，而长椅则已经成为了人们在一个欧洲小镇归属感的象征。

"荧屏"的设计体现了其作为城市中心的功能性，为大众源源不断地提供信息、转播文化节目或者体育赛事等。

该项目的灯光设计理念采用柔质防眩光面板来营造光影的效果，当灯光投射在建筑立面时，即使在夜间也能营造出强烈的空间感。在节日庆典时，可以开

启安置在广场周边的聚光灯来烘托节日的气氛。设计师除了营造树和喷泉的灯光效果，在地下停车场和车站枢纽处也采用了间接暗光技术。

新建的弗拉基广场使用的材料与相邻的圣克鲁斯广场以及附近的几个小广场、行人区使用的材料一致，既保证了材料的一致性，又给人以感官上的统一感。通过这样的处理手法，街道及其附近的绿色植物也很自然地映入弗拉基广场。

除了交通步道处，广场的铺装统一采用了蓝色的大理石，从视觉上增加了广场的面积，给人留下了开阔的印象。除此之外，这种材料的统一性也弱化了附近的移民区（位于广场北侧的葡萄牙与摩洛哥移民区）与中产阶级的高端住宅区（位于湖畔南侧）之间的差异感。拉坎布雷建筑学院的学生和弗拉基文化中心的游客造就了这一位于布鲁塞尔城市中心广场东南侧的广场的多样性，以及布鲁塞尔的都市活力。

络绎不绝的游客印证了广场所体现出的社会文化的包容性。

Redesign of a city square on top of a new underground garage and a subterranean rainwater–retention basin amidst the vivid multicultural quarter Ixelles in Brussels.

In former times there existed a lake in this place, the largest one of a chain of water surfaces. Parts of it were backfilled already in the second half of the 19th century what enabled the Place Saint—Croix in front of the church Saint—Croix and a promenade, being at that time the core of the quarter. In the late thirties of the last century, the other sides of the square were completed with the Foyer Ixellois, a complex of social housing and the National Radio building. The construction of a subterranean rainwater reservoir and an underground garage offered the chance of a redesign and reorganization of a space formerly dominated by moving and stationary vehicles.

Considering the dynamic and diversified social and urban structure of Ixelles, we saw only one solution for the transformation of this site: to create a freely accessible and generously laid out open space with no restriction, which enables and enhances co-existence; to offer a "neutral" surface for the play between solid and transparent as

well as permanent and temporary elements; to enable a multifunctional platform which doesn't limit its visitors to certain uses but motivates them and lets them act unaffectedly; to avoid a thematic characterization of the space by too strong gestures.

The challenge was to integrate the necessary traffic lines circling the square, the roofed tram and bus junction, the entrances to the underground garage and other technical infrastructure in a way, that the main function of the square as an area mainly for pedestrians, for play and rest of the residents as well as for diverse uses like the weekly fruit and vegetable market, would have absolute priority.

The conception took advantage of unique local knowledge and experience by involving the citizens into the planning process throughout all phases of design. Public consultation, workshops on various themes and acting out of scenarios allowed the residents to become "specialist consultants", whose input was integrated into the programming and development of the project, such as the choice of the mobility system, the programming of activities on the square etc. People are still committed to engaging themselves also

with the maintenance of "their" square.

Today, the sparkling water of fountains, informally planted tree groups and the choice of the species as well as structure and bluish-grey colour of the surface remind symbolically of the place's history and form a link to the neighbouring park and adjacent public spaces (Pessoa Square and Place Sainte – Croix). The "vegetation" is composed of natural and artificial elements—willows and maples connecting to the plants at the lake, willows and chestnuts to existing green spaces in the north, and the light glass and steel canopy sheltering the tram and bus junction.

With reference to the place's history, the fountains embedded into the stone surface represent one of the key elements. Their water provides an additional irregular texture above the homogeneous surface of the square and refreshing coolness on hot summer days.

Due to the fact that the paved surface of the square and the tarmac surface of adjacent streets belong to the same colour palette, the furniture represents the limiting line: The modular system of the "Flow Bench" is used at the

northwestern and southeastern side to frame the place with swinging rows of wooden seats, with backrests varying in height. They offer a certain protection against the traffic with differently combined components, and the pleasure to enjoy sun or shade where situated under trees. Moreover, by bringing together most different people of different origin the bench has become a symbol for the feeling of togetherness in a European town.

The "Screen" supports the re-instalment of central urban functions into the place by showing continuous information of the city, telecasting cultural or sports events etc.

The final realization of the light concept is on its way: it works with a soft and glare-free brightening of the facades, therewith creating a strong spatial feeling also at night. At special events, spotlights at the multifunctional masts situated at the edges of the square, can be activated. Besides illuminated trees and fountains only the entrances to the underground garage and the tramway junction are staged with indirect dark-light technique and perceived as "single actors" within the mildly lit up urban space.

Using the same material for the new public square, the adjacent Place Sainte-Croix and the surrounding smaller places and pedestrian zones, guarantees both a continuity of material and a visual continuity. Like a matter of course, streets and neighbouring green "flow" into the square.

The consistency of the surface made of blue granite, which—with the exception of the traffic lanes—covers the whole area between the facades, enlarges the place's dimensions and gives it a generous spatial impression. Moreover, this consistency seems to soften the difference between the poorer neighbourhood of the immigrants (Portuguese and Moroccan quarters north of the square) and the middle-class side (wealthy residential zones at the southern edge of the lakes). Students of the Architecture School La Cambre and visitors of the Flagey Cultural Centre contribute to the heterogeneity of this quarter south-east of Brussels' city centre and its cosmopolitan vibrancy.

The liveliness of the new place shows that Flagey's socio-cultural mosaic has found a space to express itself.

加拿大文化博物馆广场

Canadian Museum of Civilization Plaza

撰文：Claude Cormier Architects Paysagistes Inc.

图片提供：Claude Cormier Architects Paysagistes Inc. Michel Boulianne Tom Bean

翻译：谷晓瑞

"都市草原"是大平原地区的一片草地，彰显着和谐、勇气和创造力。这些文化元素与加拿大景观中的自然元素以及蕴藏于博物馆道格拉斯主教大楼的自然元素同样重要。种植着各种本地草种的草地拥有几分大草原的韵味，又与大草原有着细微的差别，那就是人工与天然之间的博弈。草丛是四季常青的，只有其间的十几棵低矮的松树向人们昭示着春去秋来。该项目为打造自然与文化相融合的博物馆拉开了序幕，设计充分利用其地形特点，根据周围环境中建筑风格，勾勒出一幅加拿大西部绵延起伏的草原景象。

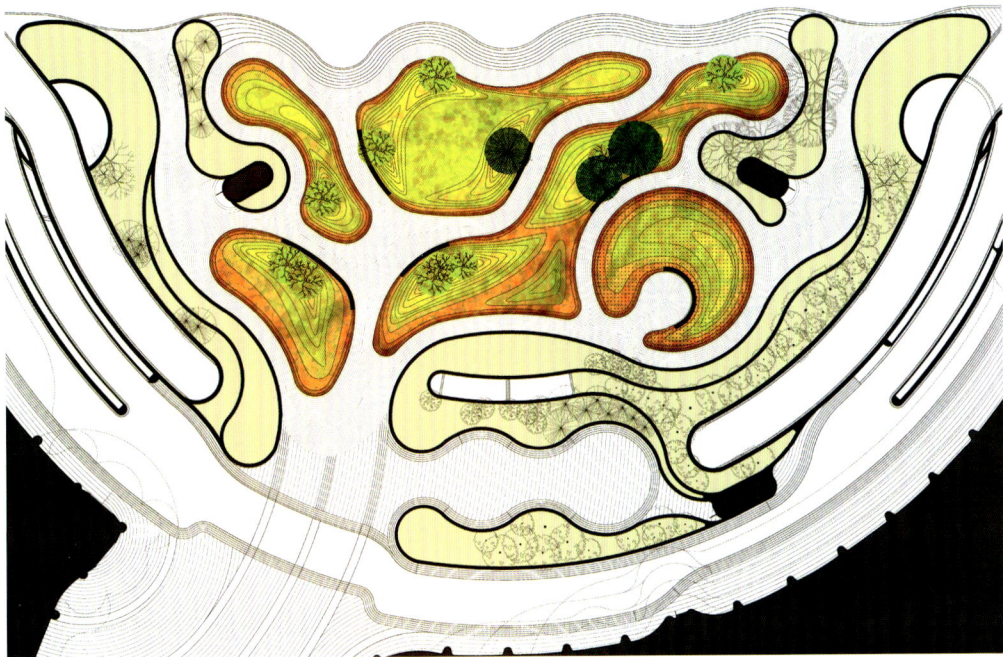

　　加拿大文化博物馆两座建筑的设计传承了加拿大景观设计的经典。博物馆大楼代表冰川，而博物馆侧楼则象征加拿大地盾。两座标志性建筑之间原来是一片光秃秃的广场，使两座建筑与河对岸的国会大楼都显得非常雄伟壮观，景象非常适合用做明信片的画面。然而，由于广场上没有行人驻足和休息的设施，这里一直很空旷。博物馆试图通过引入景观设计来解决现状，景观设计师抓住了这个机会，巧妙地设计出了和两座大楼一样具有象征意义的景观经典之作——大草原。

　　为了建造一片适合于城市的草原，设计师在硬质路面上添置了五个人工山丘，形成了这个都市草原。由于广场地下是停车场，合理的重量分布是首先需要考虑的问题。为了减轻起伏的地形所带来的压力，设计师采用轻质土，并用泡沫板来搭建山丘内部，树木则种植在地下车库的支柱地方。

　　在设计该项目之前，广场上没有任何遮挡物，经受着酷暑严寒的考验，不仅行人无法在此休息，就连周围主题花园里的植物也饱经风霜。另外，广场上也没有引人入胜的景致，所以这里一直十分冷清。

　　为了解决这个问题，设计师在山丘上横向种植了6种本地草种，随着季节的变化分别呈现出绿色、黄色、红色，草地和具有冰川特点的博物馆大楼使这里的微气候得到了改善。五彩缤纷的本地野花点缀在草原上——银莲花在春天里绽放；夏秋季节，西部红百合、天人菊、Smooth Blue Aster、管香蜂草、鼠尾草争奇斗艳。几棵矗立的唐棣和松树与横向延展的草原和建筑形成了鲜明的对比，增加了立体感。

　　加拿大大草原的历史是一个关于和谐、勇气和创造力的故事。"都市草原"这个具有象征意义的非凡创意，力求在这片充满文化气息的博物馆中给公众带来另一番体验。

Urban Prairie is a field of grasses for the Great Plains, demonstrating order, courage, and imagination. These forces of civilization are no less significant than their natural counterparts that have shaped the Canadian landscape, embodied in Douglas Cardinal's buildings for the museum. A field of mixed indigenous grasses evokes the prairie phenomena of horizon and nuance, and expresses the tension between the natural and the artificial. Embedded in the unchanging hue of grasses are a dozen dwarf pines that register the passing of the seasons. The project is a spirited overture to a museum dedicated to the merging of nature and culture. In connection with the architectural leitmotiv, the project plays with topography, illustrating the unevenness of the prairie landscape of western Canada. Contour lines establish the undulating circulation pattern.

The two buildings that comprise the Canadian Museum of Civilization were designed to embody signature features of the Canadian landscape—the Museum Building a reference to glaciers, and the Curatorial Wing to the Canadian Shield. The formerly barren hardscape Plaza that connected these two iconic buildings made for successful postcards, with its position and scale designed to optimize the view of the two buildings and the Parliament Buildings across the river. However, lack of pedestrian comfort kept the plaza empty. The Museum sought a landscape solution, which we considered as an opportunity to expand upon the conceptual metaphor of the buildings by creating a reference to another signature feature of the national landscape—the Prairie.

Adapting the prairie phenomenon to the city, we created Urban Prairie, consisting of five earthwork insertions that are embedded into a restored hard-paved surface. Situated on the existing roof structure of a parking garage below, weight distribution was a major consideration. The load of the undulating landforms was reduced through a lightweight soil medium, and mound heights were built up with panels of Styrofoam. Point loads under trees were deliberately positioned over existing columns of the structure below.

Formerly, the Plaza was exposed to extremes of summer sun and heat, and winter wind and cold. In addition to pedestrian discomfort, lack of protection against climatic extremes increased stress levels in the plants of the adjacent thematic gardens. The absence of significant features and amenities was another shortcoming that kept the Plaza bereft of visitors.

In response, the microclimate was modulated through landforms shaped to integrate the geofluvial character of the museum buildings, with six distinct native prairie grasses planted in horizontal bands along the contours of the mound topography to create dynamic striated motifs of green, yellow, and red that changes in hue and texture across the seasons. Points of colour emerge from the native prairie wildflowers distributed among the grasses—Pasque Flower debut in spring; Western Red Lily, Blanketflower, Smooth Blue Aster, Wild Bergamot, and Prairie Sage throughout the summer and fall. A few Serviceberry and Pine give vertical contrast to the horizontal emphasis of the landforms and buildings.

The history of the Canadian Prairies is a story about order, courage, and imagination. Urban Prairie aims to manifest the terrain onto which these forces of civilization unfolded, establishing this important and symbolic threshold as a comprehensive part of the public's museum experience.

奥克森富特古城中心改建

Renovation of the Old Town Center of Ochsenfurt

撰文：Valentien+Valentien　　图片提供：Valentien+Valentien Michael Latz Peter Wesselowsky　　翻译：丁岩

1

1　市政厅和集市

2　喷泉

规划平面图

对于沿美因河散步的游客来说，奥克森富特的弗朗科尼亚镇以其充满魅力的古城镇中心以及保存完好的木屋而具有吸引力。小镇市场及与其毗邻的主街和 Brückenstrasse 大街是主要的购物区，也是当地的商务和行政中心，并且分布着很多居民区。

过去，该中心区域的环境和生活方式曾一度受到繁重的交通及机动车临时限制措施的影响。因此，通过建造支路来缓解古城的交通是重新塑造古城和提升其价值的前提，同时还可以减少古城中心的交通量。为了重新构建古城，政府邀请 8 家建筑设计事务所参加了一场设计竞赛，此项竞赛的主要目的除了要将古城中心打造成为环境优美的居住区和商业区之外，更重要的是重塑这一传统的古城中心，并借此改善居民的生活方式。

Valentien+Valentien 景观设计公司于 2004 年赢得了该项比赛，并于 2005 年应邀开发和改建古城。评审团被其致力于维护古城的理念所打动，包括使用高质量的天然石块铺设人行路面，试图在市场北区营造开阔的公共场所，以及在市场南区建造钢化玻璃的亭子，即所谓的 "橘子园"。

在接下来的有市民参与的设计过程中，最初的参赛设计发生了些许改变——"橘子园" 变成了喷泉，设计师对 St. Andreas 教堂的古墙进行检测后发现其缺乏稳定性，最终决定将其全部重建。

奥克森富特古城的市场和 Brückens-trasse 大街将采取措施减少交通流量，从而提高市民及游客的生活质量。路面的设计不需经过明显的改动就可将这一区域变成步行街。

简洁典雅的砖石路面在衬托出宏伟建筑的同时，又不显得突兀；用花岗岩铺设成的古典的砖石路与现有的小径相映成趣，路面的质感和外观都与古城相衬，而且结实耐用；主街和 Brückenstrasse 街的机动车道以碎石按照人字形铺设而成；石板铺成的下陷明渠排水系统，与狭窄的街道相呼应。

相反，公共区域包括一条不对称铺装的石质人行道，与北侧建筑相互平衡，同时还可以作为排水系统。商店门前原有的人行道被拓宽，拓宽后的户外区域可暂时性或永久性地作为咖啡馆或啤酒花园。设计师还针对户外家具和太阳伞提出了一些建议，使商铺可以通过户外装饰达到形象上的统一协调。

新市政厅前的公共广场以宁静作为设计的基本要求。设计师将 4 棵树木作为十字路口的标志；北面的墙体界定了广场北侧的界限，也为游客提供了小憩的场所；将现有的斜坡进行平整，以便在广场上举行节日庆典和展销活动。

在靠近大街的市场南区将修建一个略高于地面的喷泉，狭长的水岸使人们联想到古城的名字，同时也使石质的人行步道显得清新自然。

设计师在市场南区和教堂广场的连接上考虑了很多。除了设计从教堂到市政厅的斜坡外，还设计了一个从现有台阶到喷泉广场的入口通道。许多小径都汇集在这里。

设计师将教堂广场上的墙体高度略微降低，从而使教堂的景色有所改善，并减轻墙体的厚重感。由于要改建教堂墙体，St. Georg 教堂前的石头雕塑被重新放置在一处位置显眼的浅色钢质结构上。

改造后的古城中心深受市民的喜爱。在阳光明媚的日子里，大街和广场上热闹非凡：人们沿街散步，孩子们在喷泉附近玩耍，而家长们则可以放心地坐在咖啡馆外面的长凳上休息，看着熙熙攘攘的过往行人。

For visitors along the wine route of the river Main the Lower Franconian town of Ochsenfurt is a center of attraction due to its charming historic town center and its well-preserved frame houses. The market place, together with the adjacent Main Street and the street "Brückenstrasse" is the central shopping mall, which provides business and fulfils the center of administration for local and regional areas. At the same time this area provides residential space for many citizens.

The atmosphere and lifestyle used to be strongly affected by the ongoing traffic and the provisional measures to reduce traffic. Parking and regular traffic still dominated and lead to a less enjoyable lifestyle. With the building of a bypass road the traffic in the Old Town was partially reduced. It was the precondition for the beginning of the renovation and revaluation of the Old Town. It was planned with the renovation that the inner town would have a reduced traffic zone. For its remodeling eight architecture offices have been invited to a design competition. The main goal of the competition was—besides the enhancement of the inner city as attractive residential and business site—the revaluation as a traditional center of the city and the improvement of lifestyle.

The office for landscape architecture Valentien+Valentien who won the competition in 2004 and in 2005 was asked to develop the design and construction planning. The jury was convinced by the precise concept, which was maintained throughout the Old Town, by the use of high quality natural stone as consistent paving, by the intended wide common area at the northern market place as well as by the revaluation of the southern market place by a steel-glass pavilion, the so called "Orangery".

During the ongoing design process and the civic participation, several changes have been made compared to the original design of the competition: the "Orangery" has been replaced by a fountain, and surveys of the historic stone wall north of the St. Andreas church resulted in the total renovation and complete reconstruction of the wall due to lack of stability.

The market place and the street "Brückenstrasse" of the Old Town of Ochsenfurt have been designed to reduce the traffic, so that the quality of lifestyle for the citizens and the visitors will be considerably improved. Change the area into a pedestrian zone at any time without noteworthy modifications.

Elegant paving effectively presents the monumental buildings without dominating the scene. Classic stone paving made of granite is used in accordance with the

1　石雕

2~5　古城重建前后的景象

市中心教堂前的喷泉剖面图

existing alleys. The quality and appearance of the paving is appropriate for the historic areas as well as sturdy and durable. The lanes in the Main Street and the street "Brückenstrasse" will be marked by a cobble stone pavement set in a herringbone pattern. Open channels built from depressed slab stones work as drainage system. These channels also display the narrow street spaces.

In contrast the public places will have an asymmetric profile with a slab stone pavement set in a row. Public places will be subdivided by a broad pavement which is also a drainage system that runs parallel to the northern buildings. The original sidewalk in front of the shops will become a considerably wider zone, which is sufficient for outdoor gastronomy. There will be the possibility to use the outside area as a café or beer garden—temporarily or long term. Suggestions for certain types of outdoor furniture and sunshades were made. This furniture should be used by restaurant owners and other businesses, in order to obtain a consistent image.

The public square in front of the new town hall is designed in a tranquil way. Four trees will mark the cross. A wall marks the northern border and invites to have a break. The existing slope will be leveled so the area will provide space for festivities and markets.

The southern market place next to the street area obtains a slightly elevated fountain place. A long and narrow band of water reminds of the name of the town. It refreshes the stony pedestrian area. This area is connected to a protected green courtyard.

Much attention has been paid to improve the connection between the southern market place and the church square. In addition to the ramp, which leads from the church towards town hall, is an entrance to be designed from the landing of the existing stairs to the fountain square. This new opening will generate various foot-paths.

The wall on the church square will be slightly reduced in height, so the view of the church will improve and the heaviness of the wall will be reduced. The stone statue of the Holy St. Georg had to be removed due to the renovation of the church wall. It will now be positioned on a light steel construction in a new prominent location.

The inner city enjoys great popularity from their inhabitants. On warm, sunny days, streets and squares will be lively inhabited, people will stroll along the streets, stop at a café for quick refreshment, and children will play in the fountain while their grandparents rest on the benches to observe the hustle and the bustle.

3

Afrikaanderplein 广场

Afrikaanderplein Square

撰文：OKRA landschapsarchitecten　　图片提供：Benter Mull　　翻译：王颖

自 1999 年鹿特丹市议会成立基金会以重新设计 Afrikaanderplein 广场以后，Feijenoord Urban District 便积极地参与到该项目的发展规划中来，很多当地团体和利益相关者也纷纷组织会议来表达他们的意愿。

为了最有效地使用基金，设计师们首先制定出了一项总体规划。该规划的核心就是发挥广场的公共职能，以实现广场的开放性和公园的实用性。该项目的特色一方面体现在其作为广场所拥有的较高的绿化覆盖率，另一方面则体现在多条街道汇聚于此而产生的多样性文化。由特定功能框架围绕的中心区域是实现公园功能性的基础。总体规划以可供游客免费使用的中央区域为基础，而大面积的开放性区域则是公园的真正核心——人们可以在这里散步、踢球或是野餐等。各种功能设施围绕在广场周边，这些设施需要有特定的布局，如市场、游乐场以及大型鸟舍等，分三面环绕在广场四周，只留下一侧静谧的空间，植物园和清真寺便坐落在这片宁静的绿洲之中。

架有单桥的水景确保了这里的宁静。平整的草坪和交错的路径构成了公园的心脏，游客在这里可以随心所欲地漫步、休憩、踢球、野餐，或是骑马、喂鸭子等。

广场的公共草地是公园中最开阔、最青翠的区域。一道凸起的楔形台阶将公共草地一分为二，楔形台阶的一侧是舞台，由纵横交错的线围绕而成。步道由天然石头铺设，为外部的特定功能形成了一个过渡区域。外部区域以茂盛粗壮的列植梧桐树形成轮廓，作为该项目与周围街道的过渡。公园大门不但是公园的核心，而且保证了公园的功能。每天清晨，公园大门缓缓滑开，仿佛是一道透明的墙，阻隔公园与外部世界的联系。

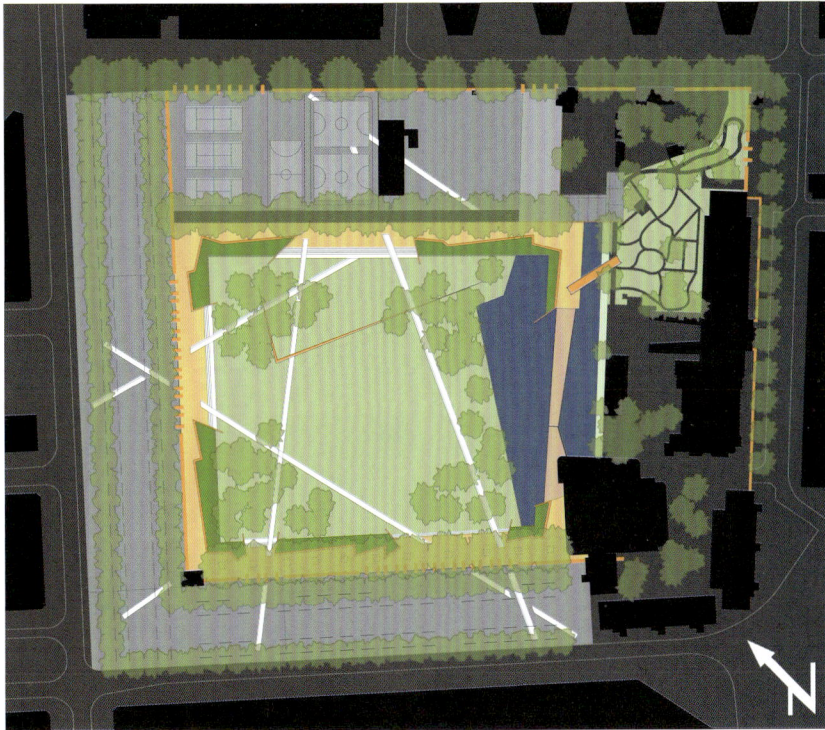

speelzone

omloop met zitmuur

trap

scheg

weide met paden

markt

boomgroep

vijver

vlonderpad

brug

botanische tuin

vogel opvang

After the city council made funds available at the start of 1999 for the redesign of the Afrikaanderplein, Feijenoord Urban District has been intensively involved in the planning development. Many meetings were initially held with residents' groups and interest groups in order to catalogue their wishes, and OKRA was commissioned to draw up a design.

In order to use the available funding as efficiently as possible, a master plan for the Afrikaanderplein was first drawn up. Central to the master plan are the public function, accessibility and park usage. The Afrikaanderplein is characterised on the one hand by greenery—the park—and on the other by the diversity of cultures in the neighbourhood which meet here. It is the basis for the functional approach of the park; a free central area surrounded by a framework of specific functions. The master plan takes as its basis a central area which is freely available for use; the true core of the park, where it is possible to walk, play football, picnic, etc., in a large open space. The "set" functions are grouped around this area, which require a specific layout, such as

the market, playground and aviary. Three sides of this zone are used a great deal, and one side is quiet. This is the green oasis, where the botanical gardens and mosque are located.

A water feature with a single bridge ensures that this area stays quiet. The heart of the park is formed by the parkland and the orbital pathway, an area for unrestricted park use, strolling, sitting, kicking a ball, picnicking, horsing around, feeding ducks, etc.

The parkland is the most open and green area. A raised wedge bisects the parkland. On the edge of the wedge is the stage, which is surrounded by diagonal lines, which criss-cross one another. The orbital path is paved in strips of raw natural stone, and forms a transitional area to the specific functions in the outer skin. The outer skin, with a roof of plane trees, forms the transition to the city life in the streets surrounding the Afrikaanderplein. The park is shut off from the outside world by a park gate, which delineates the core of the park and ensures quality use of the park. In the morning, the plates of the gates are slid open along a great length, so that it acts as a transparent wall.

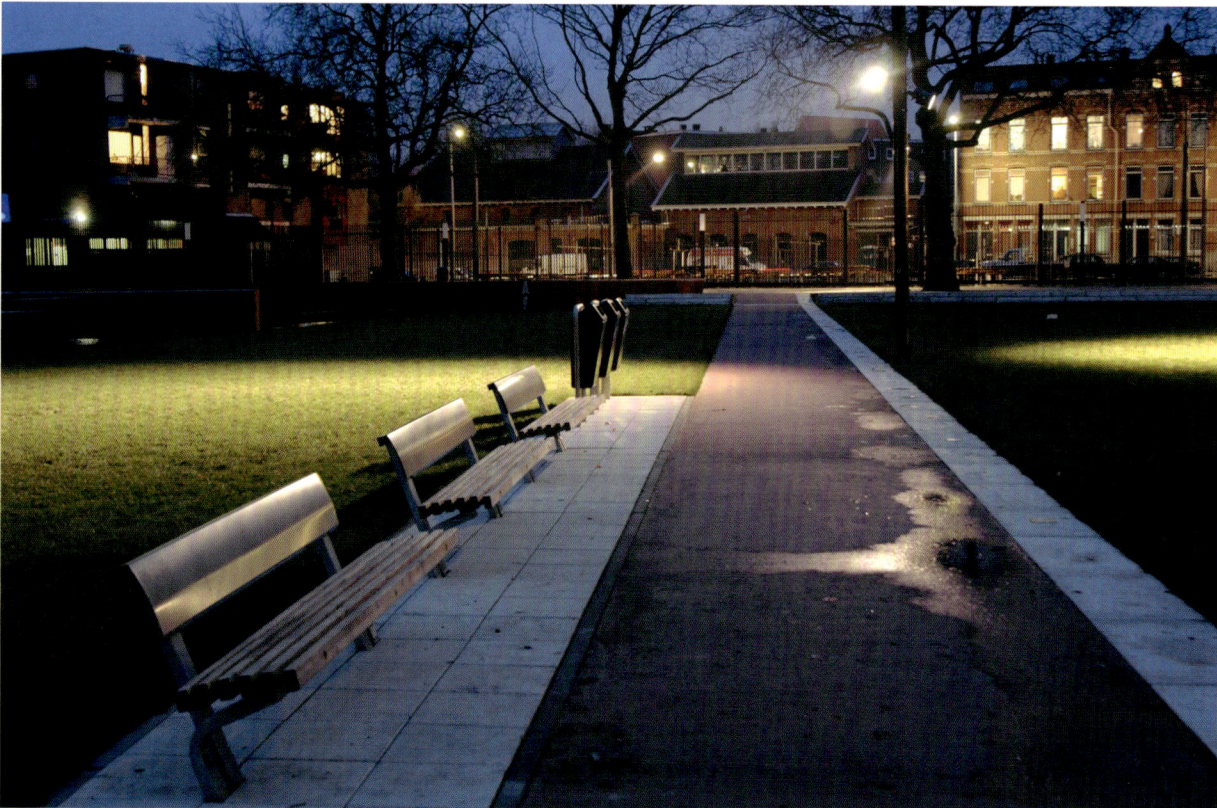

老市场广场

Old Market Square

撰文 / 图片提供：Gustfson　　翻译：刘宏阳

该项目位于诺丁汉市，是现代都市中一处安全、和谐的空间，备受广大市民的喜爱。人们从冗繁的日常生活中抽身来到此处，等候亲朋、约见好友，不仅能够获得片刻的宁静，还能欣赏到精心编排的城市文化演出，获得愉快的身心体验。

该项目是英国历史上最为悠久的公共广场之一，拥有 800 多年的历史，曾作为市场使用。该项目占地面积为 11 500m²，是仅次于特拉法尔加广场（Trafalgar Square）的第二大广场。1920 年，设计师 T.C. 哈威特对该场地进行了设计，但随着城市的发展，老广场已经无法满足诺丁汉市作为欧洲大都市的发展要求，广场改造势在必行。2004 年，一场国际设计竞标大赛拉开了帷幕，设计师们展开了激烈的角逐，最终评委会一致决定依据新方案的设计对广场进行改造。

该项目具有良好的通达性，选用高品质的材料创造出全新的充满动感的水景，并且运用软质景观，与街头设施相互融合、协调自然。该项目全天 24 小时开放，其设计本身充满了亲和力，吸引着四面八方的行人到此休憩。该设计创造出一块灵活多用的表演空间，使改造前在场地周边举办的活动得以移入广场内进行，因此这里经常举办丰富多彩的市民活动。这些基本的设计理念为场地注入了一种独特的空间特质，成为了诺丁汉市一处充满个性的公共空间。

设计通过台阶来调整场地的高差，每一级台阶的颜色各不相同，由大块的花岗岩筑成。在地势上，台阶低于广场平面，其宽度呈递减趋势延伸，并且台阶两侧设有成排的坐椅、植栽容器和水景。在台阶环绕的宽敞的广场上，经常举办贸易活动和市政活动。

该项目的水景呈现梯田式的布局，包括水池、瀑布、水渠、喷灌水景和雾状水景等。这些水景交替搭配，呈现出如舞台剧般的动感效果。除灯柱外，所有的灯光设计都被设计师精心地掩藏起来，在夜晚营造出梯田式的灯光效果，灯柱还可以用做一些临时活动的支架。

该项目特别注重交通的通达性，设计师在香波酒

吧（Chapel Bar）与史密西路（Smithy Row）、朗格路（Long Row）以及费莱尔巷（通往诺丁山城堡）之间设计了全新的直达路径，保证了四面八方的人们都能够便捷地到达广场的中心，同时也保证了人们从广场可以前往城市的各个角落。

自该项目开放以来，诺丁汉市许多大型的、公众参与率极高的活动都在此举行，其中包括免费的音乐会、焰火燃放表演、滑冰活动、食品展、灯展和花市等。随着该项目的影响日渐扩大，不仅能够在午餐时间和黄昏时间吸引市民到此休憩，在周末的时候还吸引了更多的游客到此观光。人们喜欢这座广场，并对它赞誉有加，而且市议会的领导也曾表示：该项目远远超出了预期的期待。

The Old Market Square, Nottingham is the city's guardian space, a safe haven, a place to regain energy, wait and meet friends, be diverted momentarily from one's daily routines and to experience spectacular and well organized civic and cultural events.

The Square is one of Britain's oldest public squares, with an 800-year history as a market place, and at 11,500m², the second largest after Trafalgar Square. The formal 1929 design by T. C. Howitt did not serve the requirements of a progressive European city, and following an international design competition in 2004, the Jury unanimously selected the new contemporary design.

The brief was to provide unhindered access for all, use high quality materials, provide new water features, introduce soft landscaping, integrate street furniture, create flexible performance space, allow people to linger, encourage 24 hour use, enable perimeter activity to spill out into the space, and attract pedestrians by virtue of its design. It also had to create a sense of place and reinforce the distinctive qualities and character of Nottingham.

Terraces of coloured granite blocks delineate level changes and hint at the geological strata below the Squares surface. Their tapering forms accommodate rows of benches, planters and water events, set around a large flat and unobstructed surface used for markets and city events.

The terraced water feature comprises a reflecting pool, waterfall, rills, jets and a scrim that can be switched off and enables their use as an amphitheatre. All lighting is concealed, except for masts that create a range of lighting moods and support temporary trusses needed for events.

New direct, diagonal routes between Chapel Bar and Smithy Row, Long Row and Friar Lane leading to Nottingham Castle, enable access to the centre of the square and at the same time ensure pedestrians easy access to all parts of the city centre.

Since opening, the square has hosted some of the largest, and best attended events ever staged in the city. These have included free concerts, firework displays, an ice rink, fine food fair and a bulb and flower market. The day to day impact of the new square has been immediate, and it has already become a well used space at lunchtimes, and early evenings, in addition to a popular attraction for tourists at the weekend. Local public opinion has been very positive and the head of the City Council has said it has exceeded all expectations.

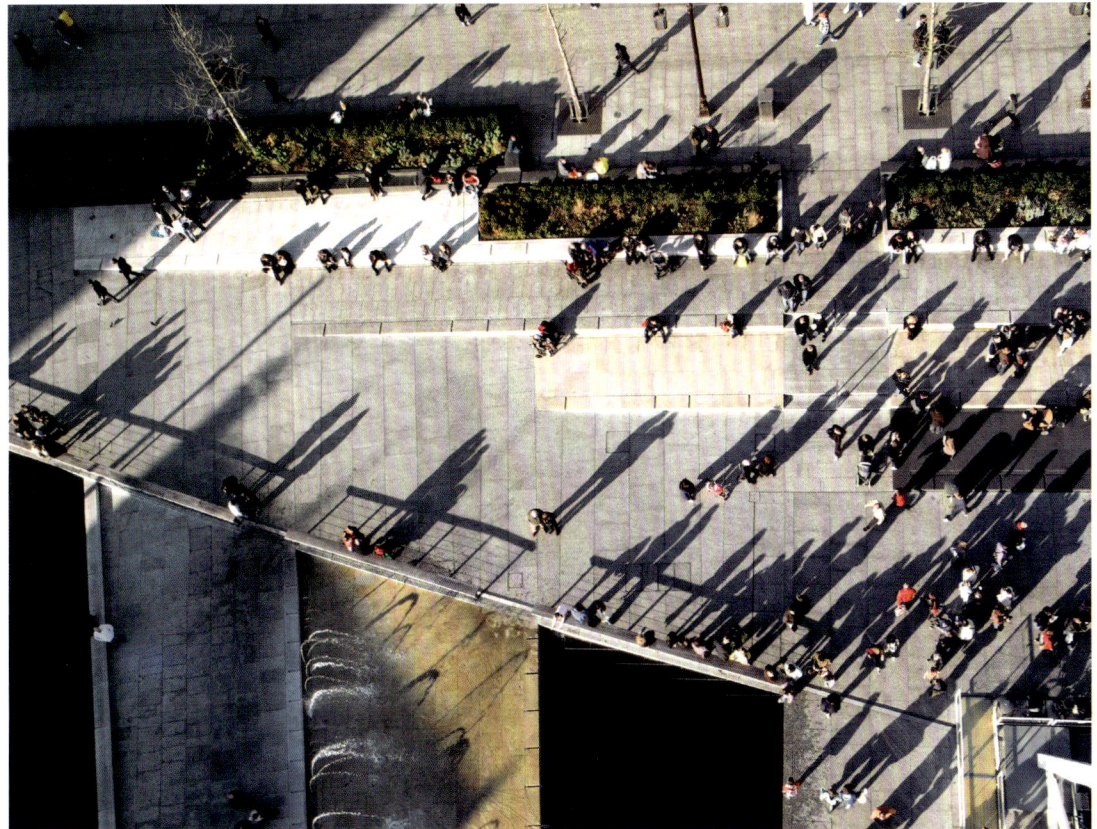

蒂罗尔州哈尔城镇广场

Town Square in Hall, Tyrol

撰文：AUBÖCK+KÁRÁSZ LANDSCAPE ARCHITECTS AND ARCHITECTS

图片提供：Martina Posch　　翻译：李博

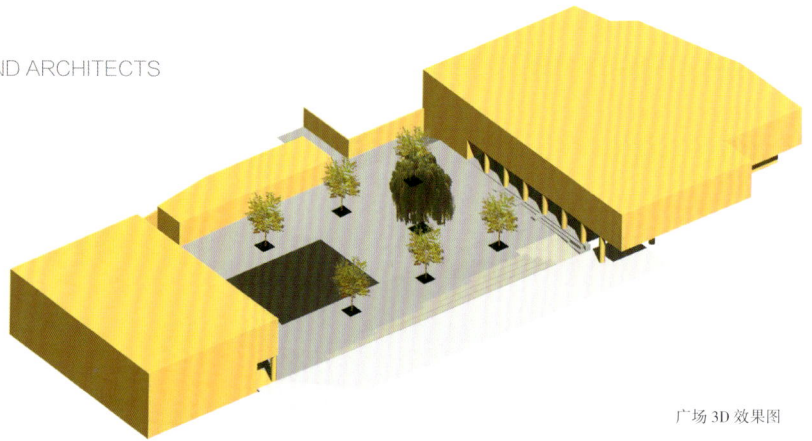

广场 3D 效果图

1 广场全景
2 俯瞰广场

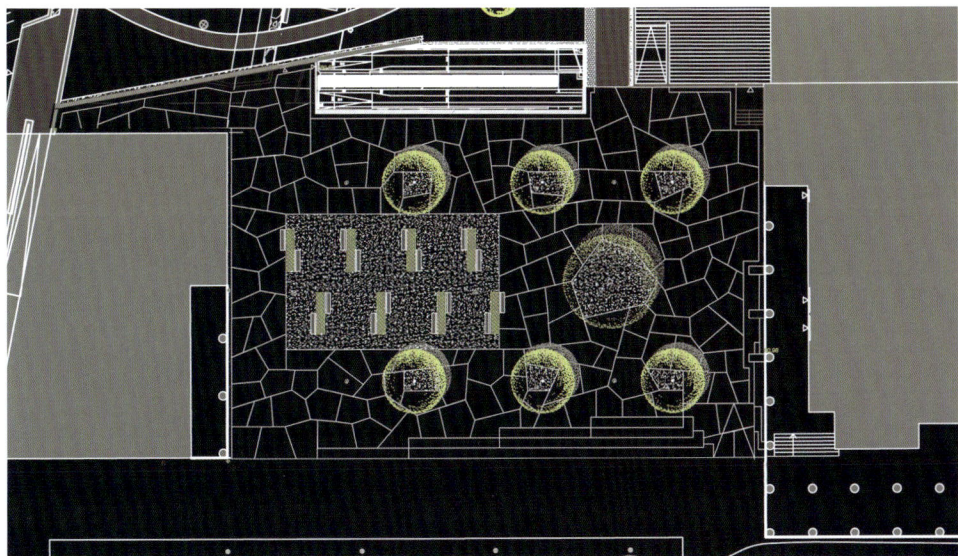

广场总平面图

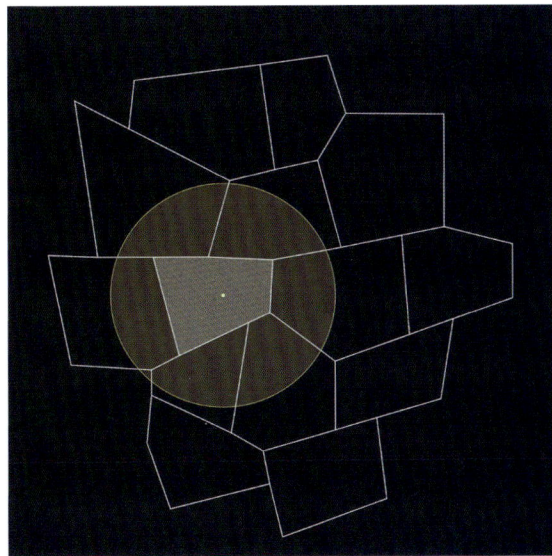

地面铺装的几何图形

起源于中世纪的哈尔镇位于奥地利蒂罗尔州阿尔卑斯山的西部，距离因斯布鲁克市 8 公里。

该项目的设计目的是为这个历史城镇边缘的两栋著名建筑和一座会议中心营造出适合的景观。广场北部的公园高出项目场地约 3 米，因此设计师将对这处倾斜场地的设计作为一项挑战。

该项目建造在一个 20 世纪 70 年代的地下车库上，这使得设计受到了地形条件的限制。如今设计师在广场的两侧建立了相邻的拱廊，新的地面采用大块的多边形混凝土地砖铺成，地砖之间的缝隙则由沙砾填补。地面的"碎冰"图案是设计师对其地下用途的暗指。银枫树栽种在古老的垂柳旁，竹园则提供了一处可以思考的空间。广场为周围居民提供了一处集会场所，也便于游客在此参观北边的房屋。

Hall is a town of medieval origin, situated in the Western Alps of Tyrol/Austria, 8 km from Innsbruck.

The task was to create on the fringe of the historic centre both an adequate context for two outstanding buildings (architects: Welzenbacher, 1932, Henke/Schreieck, 2003) and a conference centre. The design is part of a larger ensemble, which includes a park, located 3m higher north of the square.

The square is built upon an underground garage of the 1970s, which resulted in a rather unfavourable topographic situation. Now on the same level with the neighbouring arcades on two sides, the new surface is shaped by large polygonal fields of site-poured concrete with wide gravel seams. This design allows to cope with the particular challenges of the site concerning its incline. The "cracked ice" pattern provides a distinctive metaphor for its subterranean use. Silver maples stand alongside an old weeping willow. A bamboo garden offers a meditative element. The square serves as a meeting point for the surrounding habitants as well, as for tourists visiting the thorn house located on the northern side.

1　北部公园

2　远山与近景相映成趣

3　以远处的摩天大楼为背景的广场

4　竹园

校园

通往知识殿堂的白桦路 —— 贝茨大道

The Path of Birch Trees to Knowledge Palace — Bates Walk

撰文：Ricardo Dumont Nicole Gaenzler   图片提供：Sasaki 公司 Robert Benson   翻译：潘岳

1　鸟瞰图
2　石刻标志

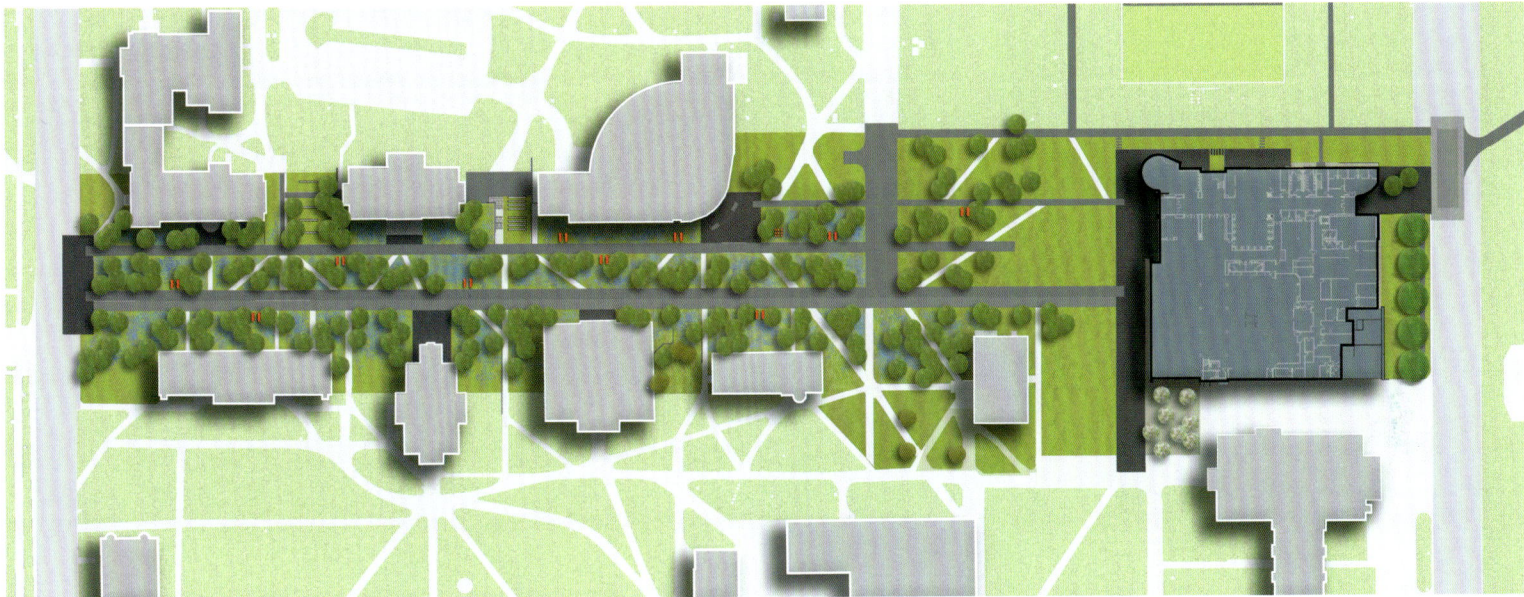

平面图

贝茨学院位于美国缅因州刘易斯顿市，是一所享有很高声望的私立学院。贝茨大道这一重要的新景观显著地提升了校园中心区的品质。

该项目既是学校的地理中心，又是举办各种活动的中心。这里曾经是一条街道和一处停车场，现已被改造成一条长约305米的、全新的、风格独特的人行步道，道路两侧种植着缅因州的特色景观——白桦树。

该项目将新餐厅和北侧主要的运动场以及住宿区相连，实际上也是连接教学区和校园社会文化中心的一条新通道。

这个令人耳目一新的标志性开放空间由一系列交错的小路组成，设计师沿小路设置了会发光的长方形坐椅，不仅照亮了小路，而且昭示着这片区域在校园中的重要地位。

该项目为促进师生进行非正式的聚会、交流想法、展开对话提供了空间。沿贝茨大道而下、朝着Andrews池塘的方向是一个露天剧场，可以进行学术讨论以及户外表演。此外，贝茨大道上还点缀着艺术品，为这一标志性区域增添了一抹亮色。

Bates College is a prestigious private school in Lewiston, ME. Bates Walk is an important new landscape element that dramatically improves the quality of the campus core.

At both the geographical center and the heart of activity on campus, Bates Walk transforms a former street and parking lot into a pedestrian landscape with an entirely new, strong identity, distinguished by a 1,000-foot-long of birch trees characteristic of the Maine landscape.

The walk links the new Dining Commons to major athletics and residential facilities to the North. Essentially, the walk provides a new connection between the academic and the social-cultural centers of campus.

The striking new signature open space consists of new pedestrian paths. The paths are lined with illuminated cube benches that light the paths and serve as beacons signifying the importance of this space on campus.

The space encourages informal gathering where students and faculty can share both ideas and conversation. An adjacent amphitheater sloping from the walk down toward Andrews Pond provides an academic forum and a space for outdoor performances. Eventually, artwork will punctuate the walk, adding another dimension to this signature space.

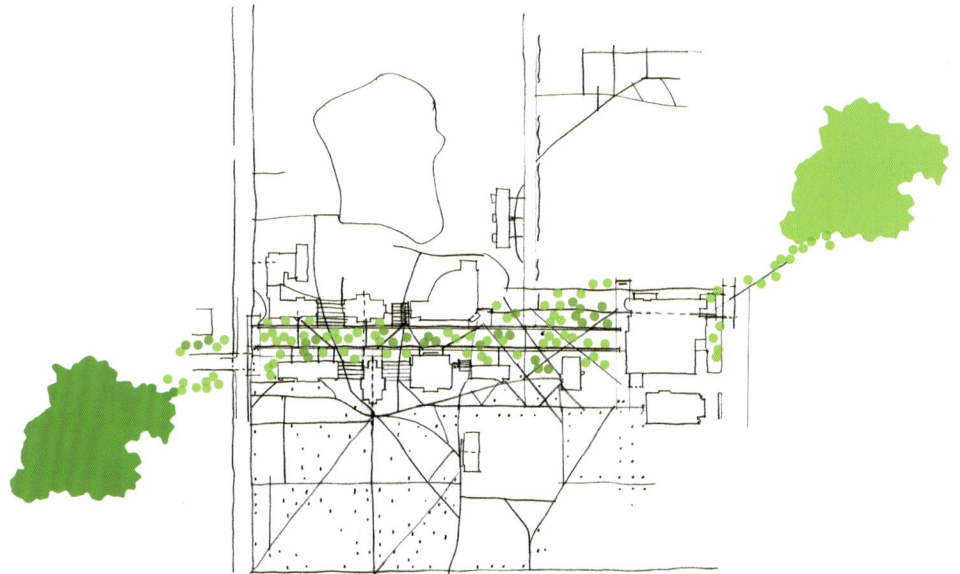

贝茨大道与原有校园小树林相接

1 聚会闲聊的场所

2 长椅夜景

3 新餐厅

慕尼黑科技大学中央庭院

Forecourt Theresienstraße of the Technical University of Munich

撰文：Valentien + Valentien Landscape Architects and Urban Planners SRL

图片提供：Valentien + Valentien Landscape Architects and Urban Planners SRL Ingrid Liebald

翻译：张璐

1

1 立于庭院中央的德国著名物理学家格奥尔格·西蒙·欧姆的塑像

2 庭院全景

慕尼黑科技大学地下停车场的顶部被重新翻修，Theresienstraße 街道与学校办公大楼之间的庭院受其影响，也得以重新设计。从街道望去，庭院在一排银杏树的掩映下若隐若现。设计师在庭院的中央设置了一处休息区，其中的坐椅由年代久远的橡木制成，师生们可以在此享受日光，也可以在树阴下休憩纳凉。

2

1　学生们在庭院中畅谈、休憩

2　金秋时节，银杏树为庭院增添了一抹金黄色

3　橡木长椅为人们提供了休憩的场所

In context of the roof renovation of the Universities' underground parking system, the forecourt of the buildings adjacent to the Theresienstraße has been redesigned. A light grove of Ginko biloba screens the forecourt from the street. Prominent seating elements made out of ancient oak logs protrude from the grove into the open court yard and invite to rest either in the sun or the shade.

总平面图

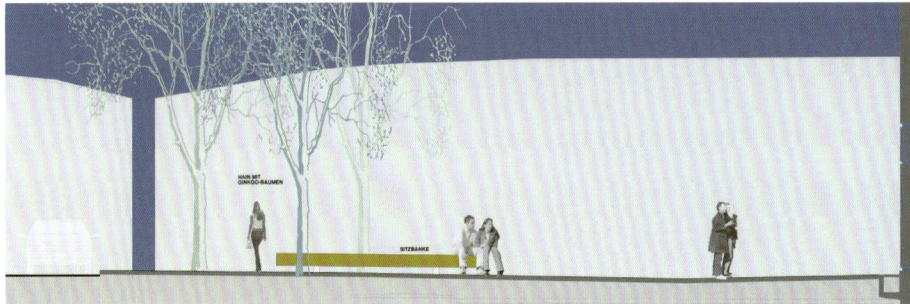

立面图

从起点到终点 —— 奥托哈恩学校的前院和操场

From Start to Finish — Forecourt and Schoolyard at Otto-Hahn-School

撰文：Katrin Klingberg　　图片提供：Marcus Bredt　ST raum a.　　翻译：刘建明

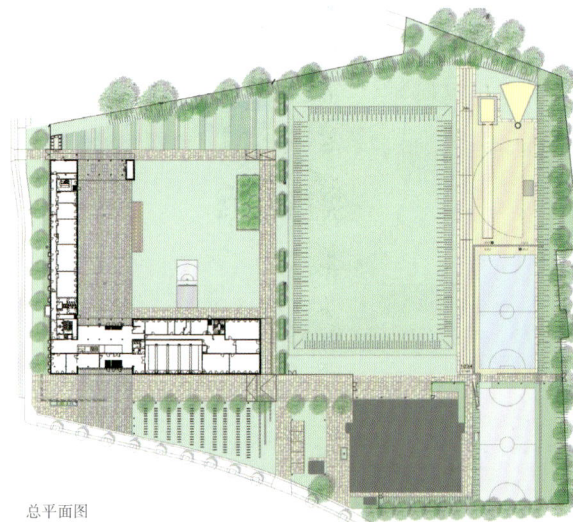

总平面图

自从求学生涯开始，学生们就有了清晰明确的长远目标。他们在以体育见长的奥托哈恩学校所接受的教育及获得的体验被艺术化地诠释到学校的设计之中。石块铺装成的"泳道"以学校的前院为起点，穿过走廊到达终点——操场；密集的铺装图案酷似涌动的水面。开放式操场可以供学生们自由嬉戏和运动——木质平台、篮球场、攀爬墙和封闭的自然花园，这些景观引导并激发出学生们无穷无尽的创意与灵感。封闭的自然花园中林木的间距非常小，生长了一段时间之后，便自然而然地形成动植物的栖息地和天然的绿色课堂。户外生物课上，学生们可以在这里观察到各种各样的昆虫。操场上还有花园和游乐区，学生们可以在这块场地中自由开展更多新奇的活动。

Pupils being thrown in at the deep end when setting out on their school careers, their education and then the experience gained during long years of schooling is artistically interpreted in the design for Otto-Hahn-Schule, a school specializing in sports. Stone swimming lanes extend from the forecourt of the school, through the lobby to the schoolyard. A densely pixeled paving pattern evokes movement through water. The open schoolyard provides space for free play and movement. A timber stage, basketball court, climbing wall and NATURE woodland stimulate the creativity of the pupils. The NATURE woodland is densely planted. Over the years it will develop into a habitat and a green classroom where species of insects can be observed during biology classes. The schoolyard leads onto the school gardens and playing fields. The school grounds will inspire interest in new activities.

充满活力的场地 —— Sct. Petri学校新运动场

Energetic Site — Sct. Petri School New Playground

撰文 / 图片提供：Lone Van Deurs Tegnestue　　翻译：刘建明

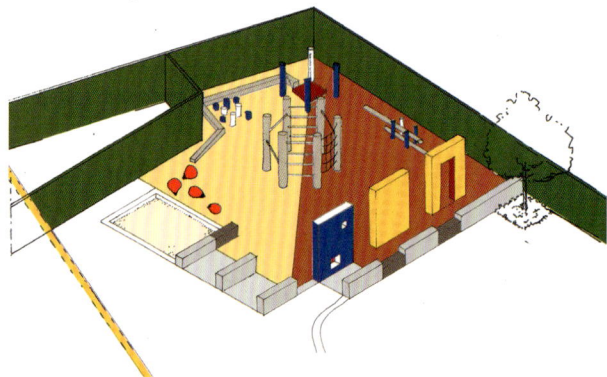

运动场的颜色搭配

Sct.Petri 学校位于哥本哈根古老的中世纪城区内，始建于 1575 年，是一所功能完备的私立学校。

在古老的红色砖墙房屋之间的校园是一处绝佳的户外活动场所。Sct. Petri 教堂附近的大树犹如绿色的幕布，遮蔽了狭窄的街道，渲染出宁静私密的氛围，体现出哥本哈根古老的中世纪城区的大部分城市景观的特征。

校方为了满足低年级与高年级学生的不同需求，有意重建校园——为低年级的学生设计一处可供玩耍的袖珍院落；为高年级的学生设计一处更为传统的沥青院落，可以进行跑步和球类运动。

校方和设计团队对袖珍院落的设计方案达成共识——不在这里安置传统运动场的体育设施，而是要建成一处能够促使学生们尽情发挥想像力和创造力的场所。设计师将曲线与正方形相结合，在场地内安置了各种材质、色彩丰富的设施，并同时采用了软质铺装和硬质铺装。铺装材料主要为沥青、花岗岩、地砖

2

1　在踏板间跳跃的孩子

2　场地中丰富的色彩和地面铺装

3　热闹的校园

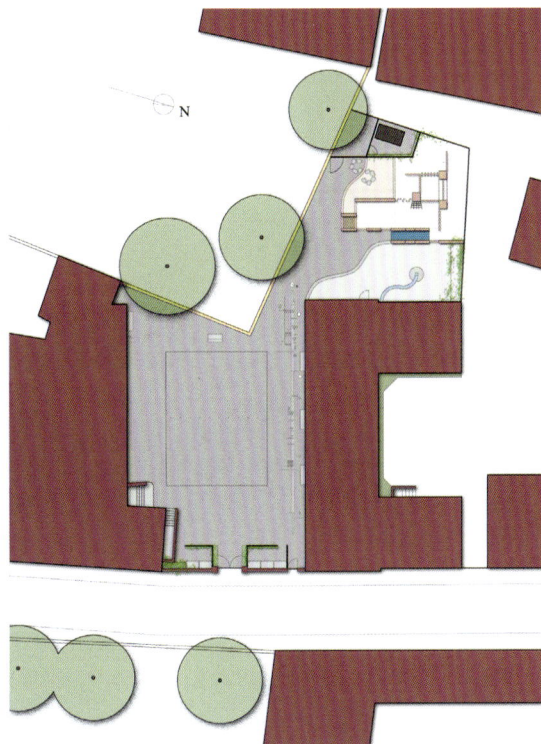

平面图

3

和软橡胶地板。距离教学楼最近的区域采用黑色地砖嵌草铺装，这样既可以观察到植物生长，也可以通过黑与绿的色彩对比给人以视觉美感；沥青铺设的区域主要是为了方便自行车的穿行，以及进行球类运动和跑步等活动；花岗岩则被设置在两种铺装材料相交的位置，在衔接区域营造出一种新的体验；在运动场放置体育设备的下方则铺设了软橡胶地板。

运动场的体育设备为孩子们提供了锻炼身体和相互交流的机会。绘有图案的墙壁营造出运动场的空间感，孩子们可以攀爬墙壁、与小伙伴们嬉戏玩耍，或是做模拟商品买卖的游戏，也可以进行球类运动。将高大的红色围栏（也用于船上防护）放置其间，孩子们可以坐在上面交谈、健身。

用于锻炼平衡能力的树桩支架分布得错落有致，孩子们可以自由地在其间穿行，作为锻炼身体的一种方式；要玩好不停晃动的踏板则需要掌握一定的技巧，经常久坐的孩子们应当适当做些这样的运动；布满石块的沙坑是小"工程师"们练习搭建房屋的绝佳场所。

如今，在哥本哈根这所历史悠久的学校里，那些可爱而充满好奇心的孩子们可以在课间或是闲暇时间里，在这个全新而现代化的游乐场中尽情地玩耍。

In the centre of the old medieval city of Copenhagen we find the German school of Sct. Petri, which is a well functioning private school for both German and Danish pupils. It was founded in 1575 but has undoubtedly gone through many changes during time.

Between the old red brick walls of the houses, lies the school yard as a nice outdoor area. The large trees at the Sct. Petri church, which is right across the narrow street, perform a green background and increases the calm, intimate atmosphere, that characterize large parts of Copenhagen's medieval city centre.

The school had a wish about a renewal of the schoolyard, which would get them a yard for the smaller children and a more traditional asphalted yard for the older children, which would have room for running and playing ball.

There was an agreement that the "the little yard", which is the yard for the smaller children, should not be filled with traditional playground equipment. Instead it should be a very special place that would appeal to the children's own imagination and creative zest. Curve lines combined with square shapes, many materials and colours plus soft and hard floorings, creates a variety of possibilities.

The coverings are asphalt, granite, tiles and soft rubber flooring. Closest to the school building are black tile with small tuft of grass. This gives an experience of growth and at the same time it creates an exciting black and green pattern. The asphalted area is for small bikes, ball play and running. Stones of granite are placed where one type of covering meets another and so the stones create new experiences around these crossings. Rubber flooring is the pad under the actual playground equipment.

The playground equipment is to appeal to movement, practice of the body and social challenges. The playground itself contains several different experiences, such as sounds that changes from one place to another. In this idea all the human senses should, in a free and personal way, be activated in the breaks between classes.

Painted walls bring spacing and heights to the schoolyard. They can be climbed on and the children can play with puppets or play shopkeepers or play ball around them. Big red fenders, which are also used on boats, are placed in small groups, so the children can sit on them and talk. The fenders are also good for exercising the body.

Planks to balance on, stumps placed with little distance so the children can walk from one to another, are other ways of exercising the body. A shaky plate performs a small challenge, which can be used to shake of the still sitting from the classrooms. And a sandpit with large unhewn stones is a great place to build houses of sand for any future engineer.

And so the school of Sct. Petri, the second oldest school in Copenhagen, with all the sweet and curious children has moved into a new and modern way of looking at school breaks and spare time playing.

福尔博格科技大学景观设计

The Landscape Architecture of Friedberg Technical College

撰文：Latz+Partner　Emporis 网络编辑 Pedro F Marcelino　　图片提供：Latz+Partner　　翻译：马秀欢

福尔博格科技大学的景观设计适应了其周边不断增长的社区需求，设计了一处配备高级设施的住宅区域，并为社区人员提供了一处可充分利用的户外活动空间。

户外空间分为两部分——高地校园和低地校园。前者与入口处等高，后者与原有的人行道、自行车道相接，通向附近中学的运动场。

在高地校园的设计中，颜色和材质的选择相互统一，而低地校园在这两方面的设计却与其相反，形成鲜明的对比。一个长约90米的水池是其主要特色，清晰地界定了学校南部与外界的界线，水池旁的阶梯很宽，可以当做坐椅。

由于校园内的设施使用率较高，设计师必须进行细致的规划以同时满足老师和学生的需求。建筑与地面应形成统一的设计体系，运动场的大小应与建筑底部的面积相当，还需设计一处与校外相连的过渡区域。运动场的一部分被呈矩形的草地替代，与大楼和整个校园内的风景相呼应。

运动场的南部有一片灌木丛，增加了石墙的视觉效果，那道石墙正是划分校园内外的分界线。设计师将唐棣属植物和鸡爪枫混合栽植，它们的花和叶构成了这里独特的风景，在秋季更是成为了一个色彩斑斓的世界。灌木丛的树影遮蔽了运动场的南部，形成了学生们的休息区。

设计师在建筑北部的路面铺设碎石，并将植栽区域相应地扩大，自行车棚架处的地面也采用碎石铺设。

水池的设计融合了绿色屋顶的设计理念，象征着大自然中水的循环。地表水流经分布细密的植物根部，一部分在流动过程中蒸发，另一部分则流入水池中，然后被引至排水沟中加以处理。

自行车棚架分布在学校的各个角落，包括主入口处的北侧、低地校园的东侧，以及南北向的建筑物突出部分的下方。坐位的布局与整个场地的规划十分协调，属于"条纹图案"设计理念的一部分。

设计师在原有围墙的基础上修建围栏，并在高地校园的一侧加设了一排栏杆，延伸至水池北面，高度为1.40米，既可防止人们攀爬，又不会遮挡住视线。

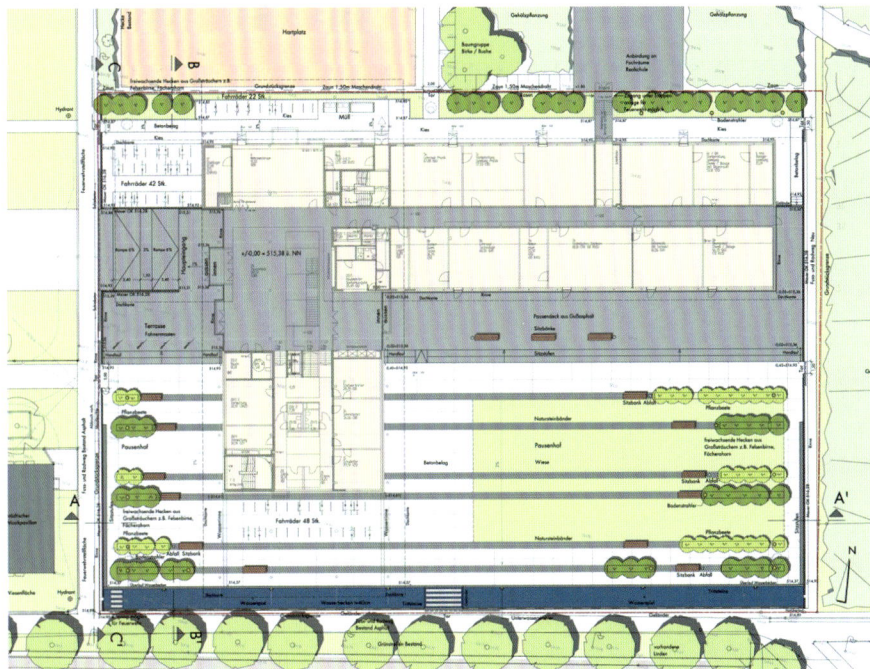

总平面图

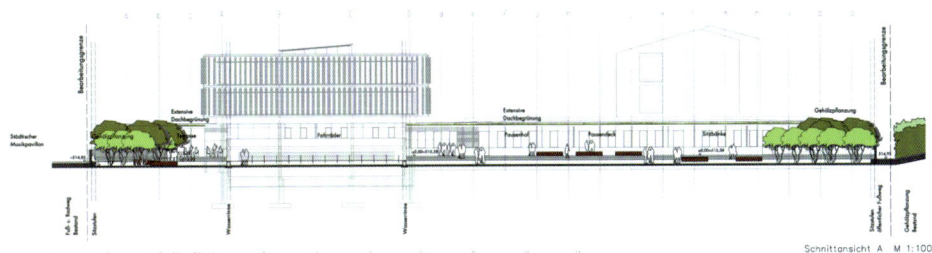

剖面图

Latz+Partner's design concept for the landscape architecture of the Friedberg Technical College (in German"Fachobersch ule Friedberg") answers to the needs of a growing community, providing a large residential district with schooling facilities for the higher levels. The design tries to provide an outdoor space that can be used by all members of the school community efficiently.

The open space is divided into a higher schoolyard level, the "yard-deck", which is at the same height as the entrance, and a lower schoolyard. The latter borders from the already existing paths for pedestrians and bicycles and to the sports grounds of the "Realschule" (High School).

The colours and materials of the surfaces in the yard-deck are consistent, whereas the schoolyard's design contrasts with the yard-deck. A water basin about 90m long is the main feature here, clearly defining the southern limits of the school grounds. The adjacent steps next to it double as seats.

Due to the intensive utilization of both yards by students and staff of the adjacent schools, there were specific constraints that included a minimum area to be paved. Building and grounds needed to form a design unit, and the size of the schoolyard had been a function of the size of the building. Clearly recognisable areas had to be created, an aspect that was closely linked with the local urban plan. These areas of the schoolyard are paved with a material that responds to the design of the building and the yard-deck, while other surfaces of the schoolyard have been partially replaced by grass. This grassed area has a rectangular shape, corresponding to the design of the building and the landscape architecture.

In the southern part of the schoolyard thin lines of big bushes were planted to extend the visual effect of the natural stonewalls marking the school area, and fencing off the public space. Juneberries (amelanchier) or Japanese maple trees (Acer palmatum) were used as multi-trunk and partly pruned wood, as their blossoms and leaves offer special characteristics, as well as an interesting colour scheme in autumn. The bushes shade the schoolyard area exposed to the south, while simultaneously defining the different lounge areas for pupils.

Inasmuch as it was possible, the surface material for the area north of the building was defined as gravel, while the planted areas were to be enlarged. Only the path to the dustbins was surfaced with the same material as the schoolyard. The soil beneath the bicycle locking-racks is also gravel.

The water basin alongside integrates a green roof section of the design concept. It also symbolizes the physical water cycle. The surface water in the extensively planted roofs and in different surfaces, on the one hand evaporates, and on the other hand passes into the water basin with some delay (due to planting), then being conducted into the infiltration ditch (once it has been cleaned).

Different bicycle locking-racks were distributed throughout the school grounds, including the area north of the main entrance, east of the low building and below the projection of the north-south building component. Generally, bicycle racks are covered. Seating arrangements are integrated into visual flow of the schoolyard and are thus part of the "stripped pattern" concept of the yard.

Fencing has been formally developed from the low retaining wall at the side of the yard-deck. A balustrade, which reaches the water basin in its northern end, was added at one side of the higher schoolyard level. The height on the outside is about 1.40 m, thus preventing people from climbing over, but still making it possible to have a look at the area.

1 草坪上的休闲长椅
2 运动场与学校的外围相毗邻

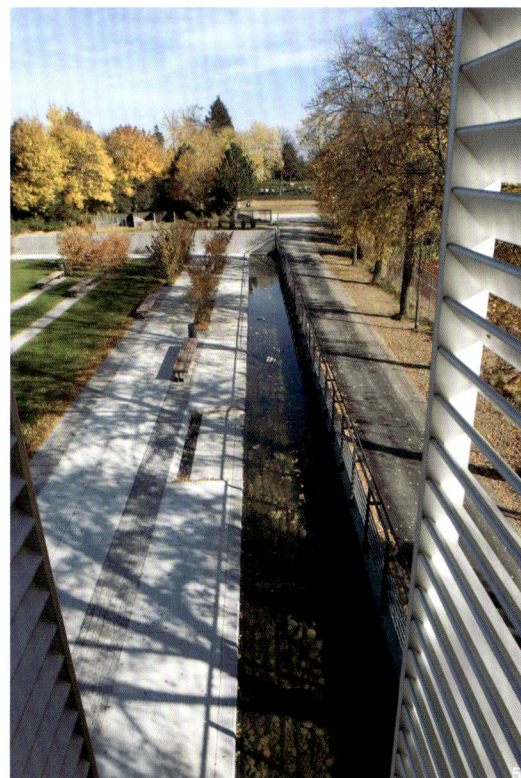

1 从高处俯瞰操场
2 坐椅与水景
3 花园水景
4 校内水景及栅栏
5 俯瞰校园内水景

屋顶空间的价值延续 —— Bakkegård校园屋顶运动场

The Extension of Roof Space — Landscape Design of Bakkegård Playground

撰文 / 图片提供：北京 KRAGH·BERGLUND 景观规划顾问有限公司　　翻译：查切尼

1　利用自然地形设计的草坪阶梯，建筑墙面独特的色块成为了建筑与一层地面的过渡
2　屋顶空间俯瞰图
3　草坪阶梯细部
4　木质的阶梯为师生们提供了一处休闲交流的开放空间

　　该项目景观设计的目的是在学校体育馆的屋顶建造一个运动场地。从高空俯瞰，整个场地的景观设计流畅贯通，犹如一条流淌于此的河流。设计师运用各种明暗色彩，使此处的景观得到突显，为师生们提供一处理想的运动空间。

　　为了最大限度地开发利用这块场地，设计还充分吸纳了教师、学生和家长的想法。当地政府的目标是在丹麦建立最完善的教育体系，使其能够不断地自我更新并具有可持续性。

　　因此，在设计中，设计师将该项目建造成了一个可以启发孩子们丰富想像力的空间。设计师在场地中设计了可供儿童游戏的沙坑、橡胶小山、大片的草坪、迷宫、木质平台、平衡木、橡树桩，并设置了玻璃纤维的十字形坐椅、金字塔形的照明设施、松木长凳等，这些独一无二的设计元素均是设计师专门为屋顶运动场打造的，既具功能性又有趣味性，体现了整体中见细节、细节又能反映整体的设计理念。设计师在屋顶的入口处设计了环形平台，将 2100m^2 的屋顶空间全部作为学校运动场使用，这对于在其中活动的学生们来说将是一个有趣而广阔的运动天地。

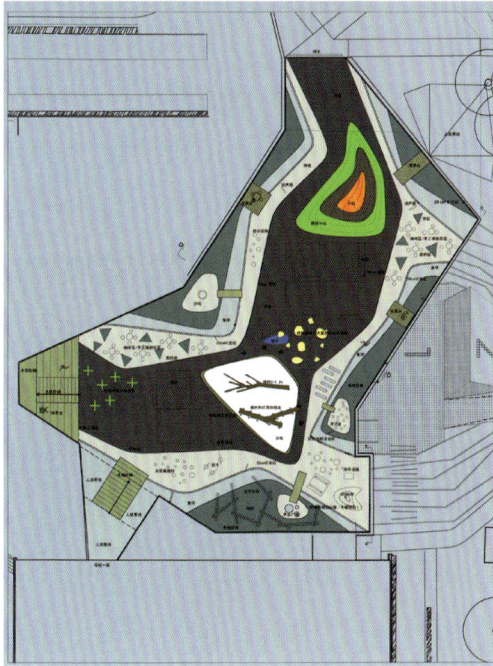

总平面图

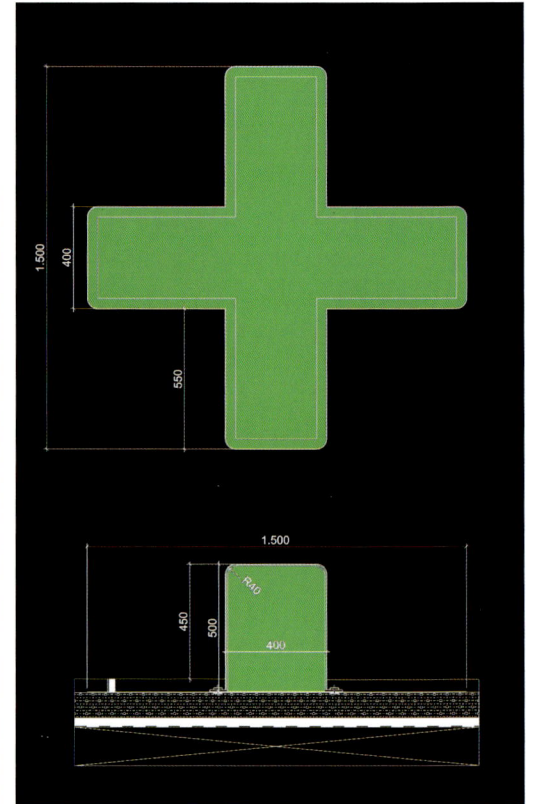

The schoolyard is constructed as a roof garden on top of the sports hall, as a relief, a floating river landscape on top of a roof, with sandpit, elephant grass, and wickerwork and rubber hills— for the enjoyment of both pupils and teachers.

The proposal is based on a value-based user involvement programme which SKUB (a school development and enlargement project in Gentofte Kommune) has formulated in co-operation with teachers, pupils and parents at Bakkegårdsskolen and in accordance with the new Primary Education Act. The vision behind SKUB is clear; Gentofte Kommune's aim is to achieve the best educational system in Denmark, an educational system which is renewing itself on an ongoing basis and where concepts that are right today will be developed and improved tomorrow.

1 内部带有照明设施的玻璃纤维十字形坐椅
2 利用自然地形设计的草坪阶梯成为师生们
   喜欢的游戏场所
3 木质的阶梯起到软性过渡高差的作用
4 从建筑通向屋顶运动场的阶梯一角
5、6 在橡树中设置的照明设施

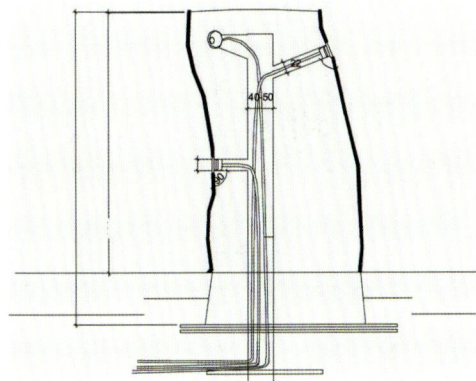

在橡树中设置的照明设施

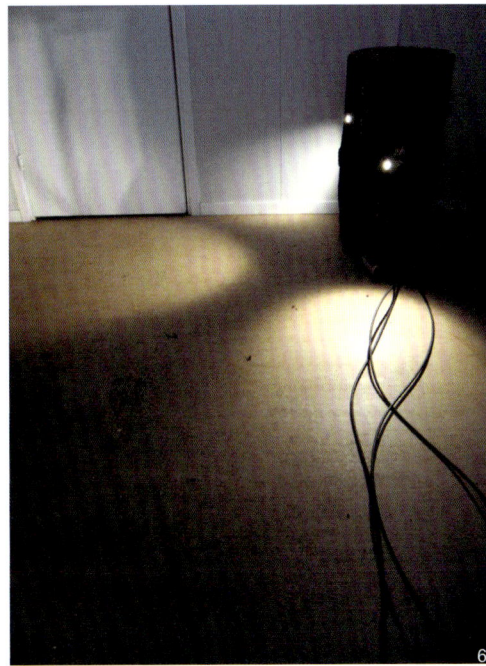

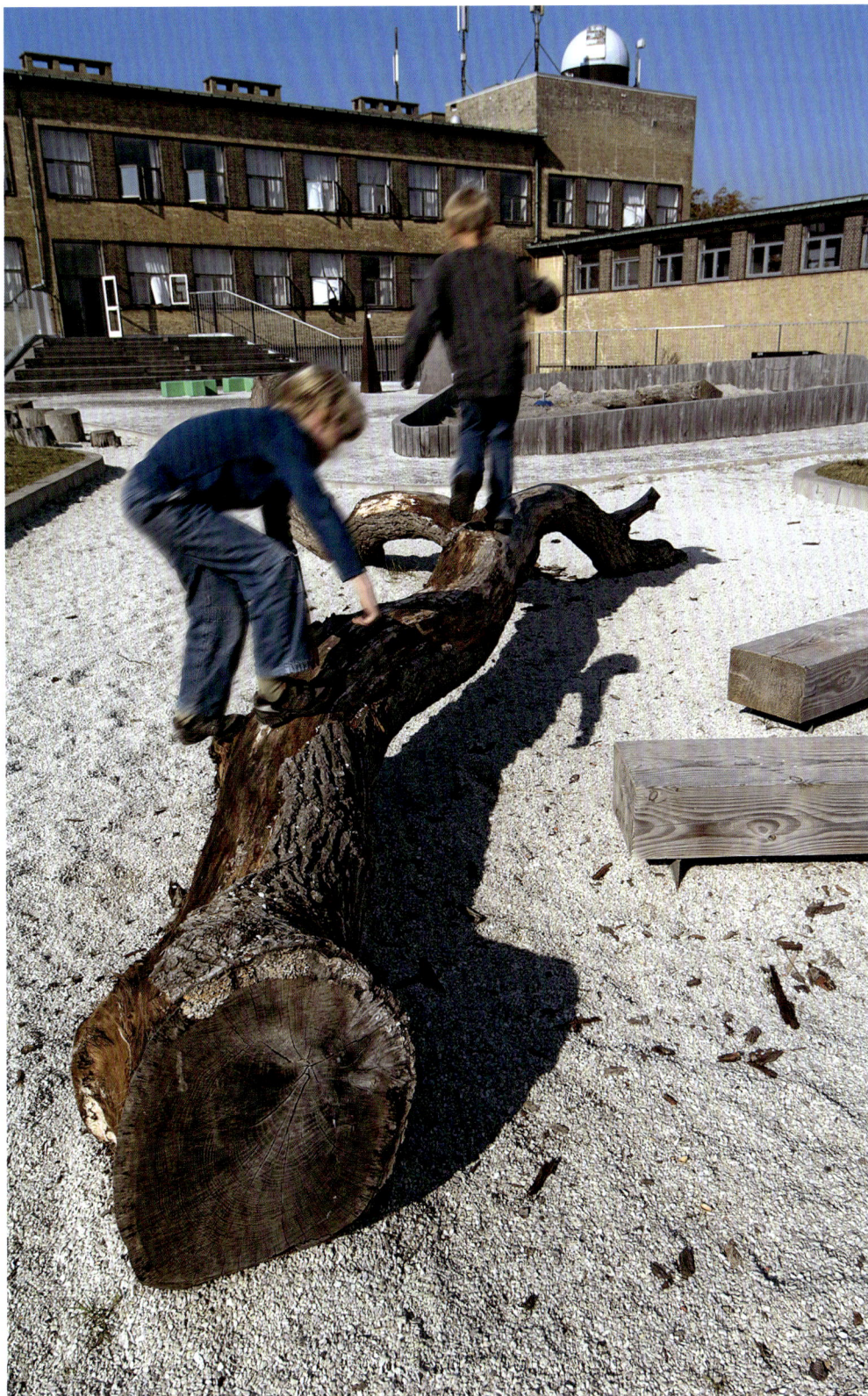

On the roof of Bakkegårdsskolen's new sports hall Kragh&Berglund has designed a playground landscape with a theme of an imaginative, floating river landscape. The landscape is constructed with sandboxes, rubber hills, elephant grass, wickerwork mazes, and wooden platforms, balancing bars, tree stumps, maze posts and trunks of oak tree. The equipment: fibre-glass cross benches, lighting, drawing pyramids and larch benches, is all specifically designed for the project. By developing design solutions for each single task together with local sub-contractors the design becomes a unique experience which links form and function. The entire project can then be seen in the detail and the detail can be seen as an integrated design solution which underlines the entire project. Encircling the entrance to the semi-sunken sports hall a giant amphitheatre is constructed with open access and with varying levels. The roof area of 2,100 m^2 is used in a new and alternative way so that the new schoolyard area becomes an exciting and different social space which stimulates variation in the children's sensuous and physical activities.

"依山村落"的互动空间 —— 巴拉曼大学

Interactive Space of a Village on a Hill — The University of Balamand

撰文：Dennis Pieprz Publo Savid　　图片提供：Sasaki 公司　　翻译：刘建明

1　新图书馆全景
2　校园入口的新喷泉

1

巴拉曼大学始建于 1988 年，正值黎巴嫩内战即将结束之时，紧邻一座已有 800 年历史的东正教修道院而建。诞生于战争岁月的巴拉曼大学肩负着一项重要的历史使命——促进中东地区基督教信徒与穆斯林教徒之间的对话。学校内的新院系也是根据需求临时设立的，并没有一个完整的远景规划。

Sasaki 公司设计的总体规划范围很广，是一个全面整合的规划，除了常规的空间总体规划外，还有全面的空间利用分析、战略性学术规划、景观规划、建筑设计导则以及融资策略。规划建议的前期建筑与景观项目，包括一栋已开始施工的新的学生宿舍，而设计的重点是校园中心区的景观和新的学校入口。

规划与设计的主要目的是营造一处能够促进对话、提高透明度、鼓励跨学科协作、倡导积极主动学习的环境。户外空间规划的重要性丝毫不逊于室内空间，必须设有从冥想步道到典礼集会场所、从小型团队活动场地到学院范围的集会场所等一系列的互动空间。

"依山村落"的理念使得校园地势较低的区域拥有足够的新空间，可容纳未来 20 年内发展规划所需增设的多数教学设施。密度是该设计的关键因素，原因有二：其一，可以促进跨领域学术思想的沟通融合，防止学

总平面图

术分裂对立，同时又可以有效地共享公共设施；其二，有助于创建一个真正的学习社区——新图书馆（即学习中心）和户外学术交流区，与新建的学生中心相邻。

精心设计的户外集会空间规模不一，频繁来往的人流为其注入了活力；多样化的景观处理手法和精心挑选的植被使这些户外空间更适于举行非正式的集会。景观是教育体验中不可或缺的部分。该项目充分利用地形优势，使人们从大多数集会空间和建筑中都可以看到远处的群山和地中海，校园设计受自然环境的启发，彰显了自然环境的特色。

设计师按照"求学之路"的理念来设计校园——这条路从山顶羊舍（社区中心）起，依次穿过橄榄林和校园，通往修道院。沿途有很多适于沉思、集会、表演、休憩和交谈的场所，还有或动或静的水景。由于沿路的部分区域集中了很多教学活动，因此这些路段很"城市化"；但是沿路大部分地区仍然是乡野风貌，为以后的其他教学活动预留了空间。"求学之路"将几个总体规划的重要目标完美地整合在一起——对话与透明度、环境敏感度与可持续性以及场所感。

该项目的景观设计力求营造一种壮观感。实现这一想法的关键是设计一个新的校园入口，强化大学是"文化方舟"的理念。此外，校园里还设计了一系列公共空间，用于整理边界、连接重要的新建筑群，并为学院开辟一个新的有影响力的公共区域。设计策略还大量采用本土材料与工艺，在推广传统的地中海风格的同时，为校园建立乡土植物设计语汇。

新的女生宿舍位于陡峭的岩石高地上，是总体规划中第一个实施的项目。周围稀疏地分布着一些树龄较小的橡树。建筑设计的基本构想是沿基地山坡创造一系列层层跌落的造型，并由一座条形建筑相连，形成了穿越地形的山脊。

新图书馆（即学习中心）是教学综合区的门户，主入口朝南，入口处设有阶梯。建筑内部包含占据二层楼的图书馆、学校档案馆、一系列教师与学生的资源中心（包括远程卓越中心、科技中心和其他辅导空间）。建筑将一个现有的户外空间围合起来，改造成一个室内中庭，连接周围现有的建筑物与图书馆的各层。这一中庭空间设有大型餐饮和休息空间，是社交集会以及整个教学综合区内小组学习的焦点地带。

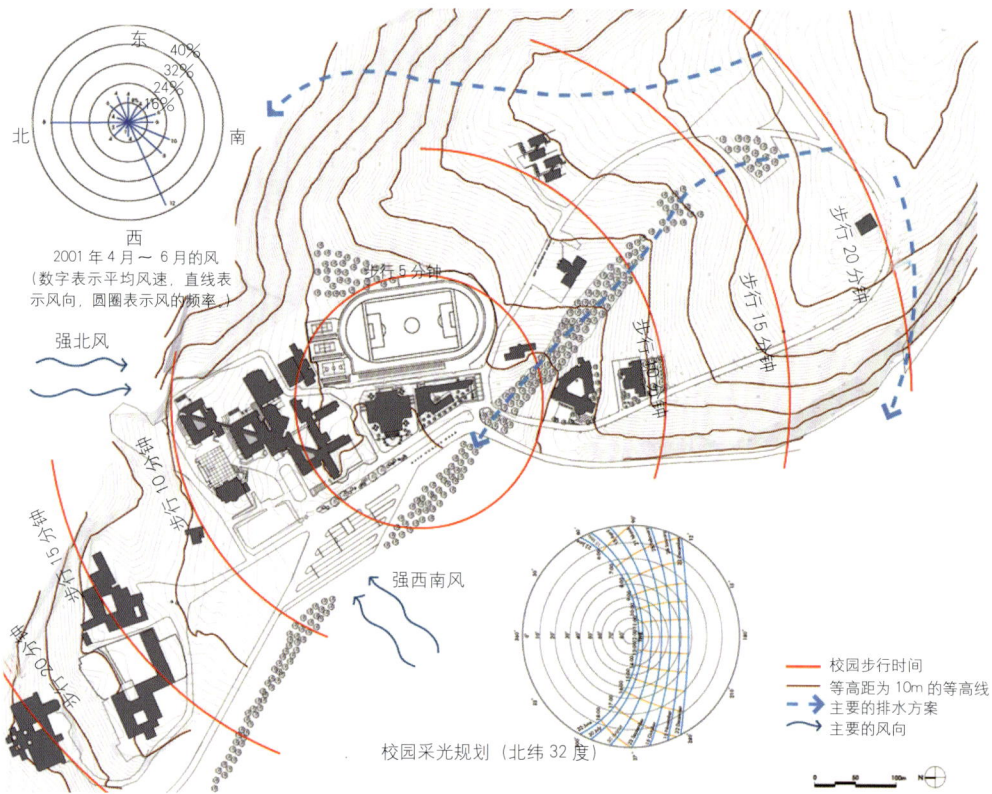

东 40%
32%
24%
16%

北

南

西

2001 年 4 月～ 6 月的风
（数字表示平均风速，直线表示风向，圆圈表示风的频率）

强北风

步行 5 分钟

步行 10 分钟

步行 15 分钟

步行 20 分钟

强西南风

校园步行时间
等高距为 10m 的等高线
主要的排水方案
主要的风向

校园采光规划（北纬 32 度）

风向、采光和排水方案研究以及步行距离

校园核心区总平面图

The University of Balamand was established next to an 800-year-old Orthodox Monastery in 1988-toward the end of the civil war in Lebanon. Critical to its mission is the commitment, born out of bitter experience, to foster dialog between Christians and Muslims in the Middle East. New faculties were created and accommodated in somewhat improvised fashion to meet specific needs, but without an overall vision.

The Sasaki Master Plan has broad reach, and is a fully integrated plan, including, in addition to the physical master plan, a comprehensive space use analysis, a strategic academic plan, landscape plan, architectural design guidelines, and fund-raising strategies. Early architectural and landscape projects recommended by the plan, including a new dormitory, have already been undertaken, with the landscape plan for the core area of the campus and a new entry playing a central role.

The central intention of the planning and design process was to create an environment that would foster dialog and transparency, encourage interdisciplinary collaboration, and promote active and engaged, learning. Outdoor spaces had to be made as significant as indoor spaces, and had to embody a full range of interactions, from the contemplative walk to the ceremonial gathering, from the small group to the university-wide assembly.

The concept of a Village on a Hill allows sufficient new space on the lower half of the campus to accommodate the great majority of projected academic growth for the next twenty years. Density is critical for two reasons. It encourages cross-fertilization of ideas, and discourages the building of balkanized academic empires, while allowing for efficient sharing of common facilities. It creates a true learning community, with a new Library/Learning Center and outdoor academic commons at its center, directly adjacent to the new Student Center.

It also makes possible the creation of well-defined outdoor gathering spaces, at a variety of scales, which would be naturally enlivened by frequency of pedestrian traffic. Varied landscape treatments and careful choice of plant materials make these spaces more congenial for informal gathering. The landscape is an essential and integral part of the educational experience. Full advantage is taken of the topography to provide distant views to the hills and the Mediterranean Sea from the majority of gathering spaces and buildings, so that the campus celebrates and is inspired by the physical environment.

The campus is shaped around the concept of the Path of Learning—a route that extends from the hilltop Goat House (community center) through the olive groves and the campus to the Monastery. Along this path are

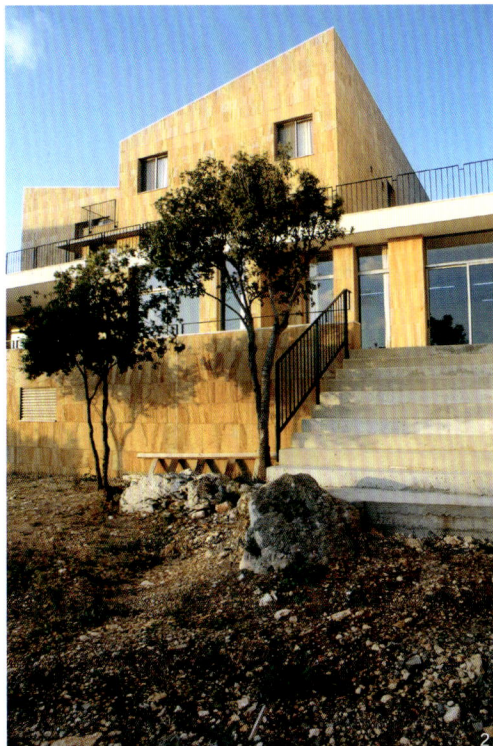

places for contemplation, places for gathering, places for performance, places for rest and conversation, and still and moving water. Part of the path is almost urban in its concentration of academic activity; much of the path is rural, leaving room for new academic possibilities. The Path of Learning weaves together several key objectives of the master plan process: dialog and transparency; environmental sensitivity and sustainability; and a sense of place.

The landscape design seeks to establish a sense of grandeur. Key to this idea is a new entrance to enhance the sense that the University is a cultural destination. In addition an array of public spaces is designed to repair edges, link critical new building groupings and establish a powerful new public realm for the university. The design strategy also capitalizes on the high level of locally available craftsmanship and materials and promotes traditional Mediterranean practices while establishing a native plant vocabulary for campus.

The new Female Residence Hall is the first building to be implemented as part of the master plan. Located along a steep rock plateau sprinkled by a young oak grove, the building is conceived as a series of forms terracing along the site slope linked by a building bar defining a ridge through the terrain.

The Library/Learning Center is conceived as a portal to the academic complex, with its main entrance oriented to the south, accessible via a series of stepped terraces. The program for the building houses a main campus library organized into three levels, the university archives, and a series of teacher and student resource centers, including the Distance Excellence Center and a Technology Center among other tutoring spaces. The new building volume encloses an existing outdoor space transforming it into an indoor atrium that links the surrounding existing buildings tot the different levels of the library program. The atrium space is inhabited by a large cafe and lounge space, providing a focal point for social gatherings as well as group study for the overall complex.

1~3　女生宿舍

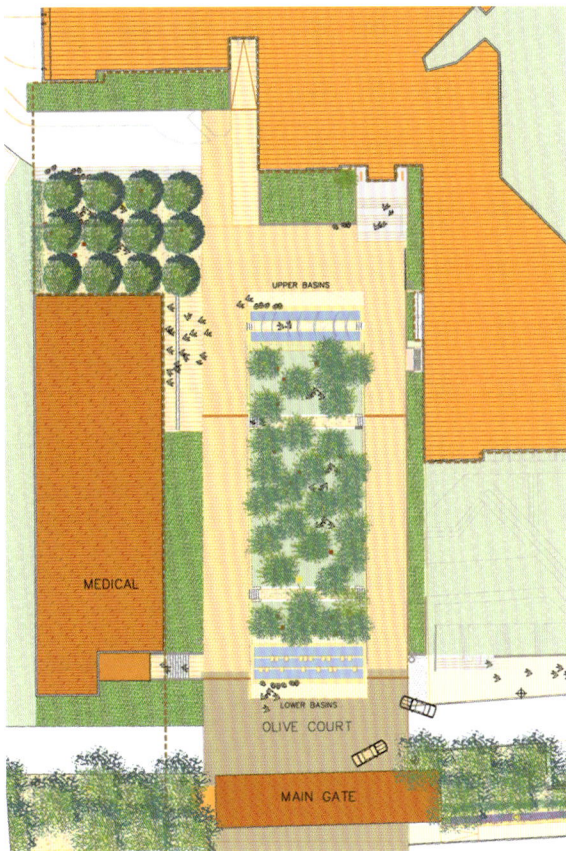

主入口和橄榄林平面图

1 庭院
2 公交车站
3 工程学院
4 新图书馆模型
5 工程学院的钟楼

女生宿舍平面图

女生宿舍立面图

商业区

Commerc

ial District

营造花园般的商业区 —— 女王购物中心

Create A Garden-like Commercial Center — Queens' Market Place

撰文：Richard L. Quinn　Robyn Sweesy　　图片提供：Richard L. Quinn　　翻译：李沐菲

1　用于表演及举办各种活动的草地和露台
2　茂盛的植物和水景使人们流连忘返

总平面图

女王购物中心位于夏威夷岛西侧的海滨胜地——威可洛亚海滩度假村,占地面积 12 500 m²,是一座多功能的购物中心。这片海滩是由附近的一座活火山喷发流下的熔岩流凝固后形成的,是世界闻名的旅游胜地之一。在这样一个独特又富饶的地方诞生了一座花园般美丽的综合购物中心——女王购物中心。

Helber Hastert & Fee 设计公司的景观设计师与夏威夷的建筑师 Ted Garduque 通力合作,共同负责该项目的植被、灌溉、照明以及硬景和水景等景观设计。该项目景观设计的初衷是在该区域营造一系列使游客流连忘返、难以忘怀的景观空间,其中包括生态池塘、富有层次的喷泉、冒着气泡的"火山"景观以及壮观的花岗岩喷泉。景观和建筑元素还包括钟表形状的雕塑、装饰性围栏、大型花园凉亭以及用于表演和举办各种活动的草地。色彩缤纷且芬芳四溢的植被与当地植被的选用,为这里的景观增色不少。

在项目实施之前,这里没有植被,只有大片的黑色熔岩,荒芜而单调,看起来就像火星表面一样。为了使这里变得草木繁茂、风景优雅,设计师从岛上的其他地方运来了大量的土壤和植物。众所周知,夏威夷海岸终年阳光充足,气候宜人。当风通过开放的熔

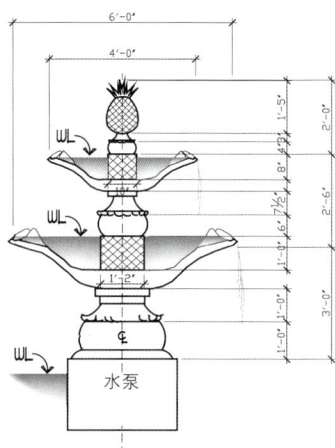

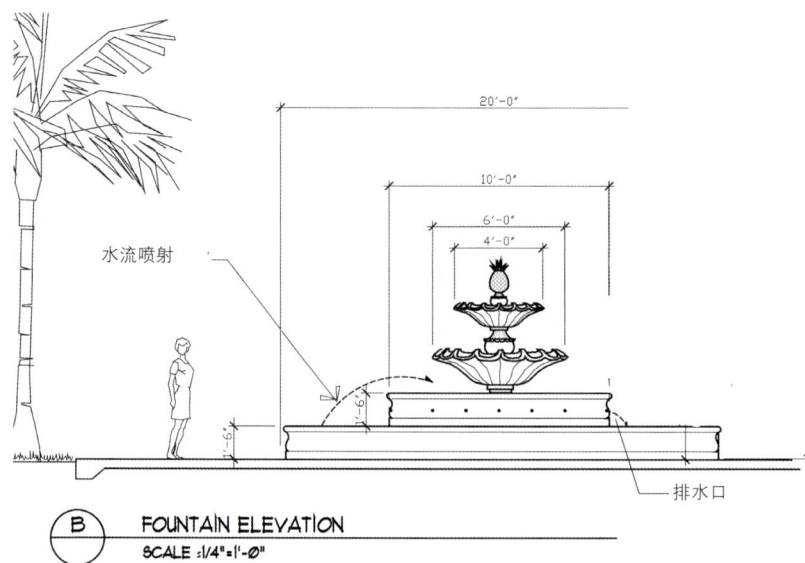

水流喷射

排水口

B FOUNTAIN ELEVATION
SCALE =1/4"=1'-0"

水泵细部图

喷泉立面图

岩区后，这里会变得干燥，设计师在设计时注意中心商业区的结构以及遮阴树木、椰子树的位置，使它们能对花园起到防护作用。该项目共使用了100多种热带植物，其中有些是夏威夷独有的本土植物，并设置了能够覆盖到所有植被的灌溉系统，以保证植物的健康生长。

　　女王购物中心的建筑及景观设计还包含一个与夏威夷历史相关的主题，记录下了皇家君主制的女王在19世纪对这座岛屿进行统治的历史。设计仿效了那一时期热带花园优雅的风格，以及夏威夷与中国通商时期的设计风格等。一些建筑特征，如塔状的屋顶设计、砖墙、装葡萄酒的巨大陶罐，都体现了当时中国在文化方面对夏威夷的影响。

1　购物中心主入口处的喷泉
2　热带植物颜色艳丽、芳香四溢
3　热带气候适合种植各种植物
4　细长的水流带着美妙的弧线喷射出来

1 花园内处处充满着色彩缤纷
  的热带植物
2 在花坛外还设计了坐椅，为
  购物的人们提供休息之地
3 池塘上的混凝土小桥
4 混凝土小桥也是花园里重要
  的景观元素

花园手绘效果图

设计中还有一个有趣的特征就是在商业中心创造性地使用了不同种类的铺装材料，包括硅岩石板、深黑色板岩石板、彩色地砖以及杂色的混凝土路面。各种不规则的图案和铺装材料使混凝土地砖具有一种舒适自然的特质，更适合于这个朴素自然的熔岩区域。

水景是该项目景观设计的重中之重，也是塑造这座花园般的购物中心的点睛之笔。富有层次的喷泉是花园内一个特别的景观元素，给人留下深刻的印象。细长的水流突然从某块岩石下面喷出来，带着美妙的弧线和不同的层次穿过小路和植物，落到地面的大陶罐和池塘里。这一景观元素非常出人意料，给购物的游客，尤其是孩子们带来了惊喜和欢乐。

购物中心的主入口处是一座传统样式的喷泉，顶端饰有石刻的菠萝，表达了对游客和购物者热烈的欢迎。喷泉的材质为花岗岩并且是在中国雕刻而成，与园内其他自然主义风格的喷泉形成了鲜明的对比，也突显了其历史性的主题。

水景的设计由 Helber Hastert & Fee 设计公司提出，并通过一系列的规划草图及平面图展示出来。水景的安装，包括水泵及过滤系统的设计都是由夏威夷的水景专家组 Kai Pono Builders 完成。为了营造美丽的池塘和近乎于天然的瀑布，设计师使用了多种建材及技术。园内的池塘上，就有一座惹人注目的"熔岩"桥，充分体现了设计者卓越的审美能力和不凡的创造力。

3

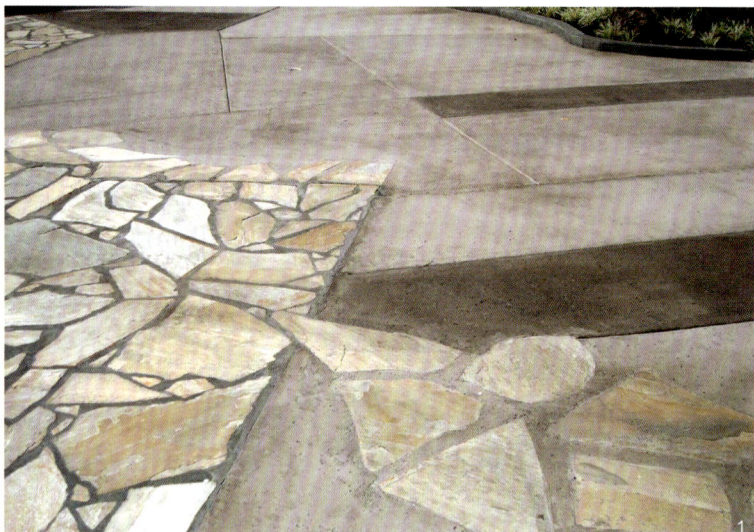

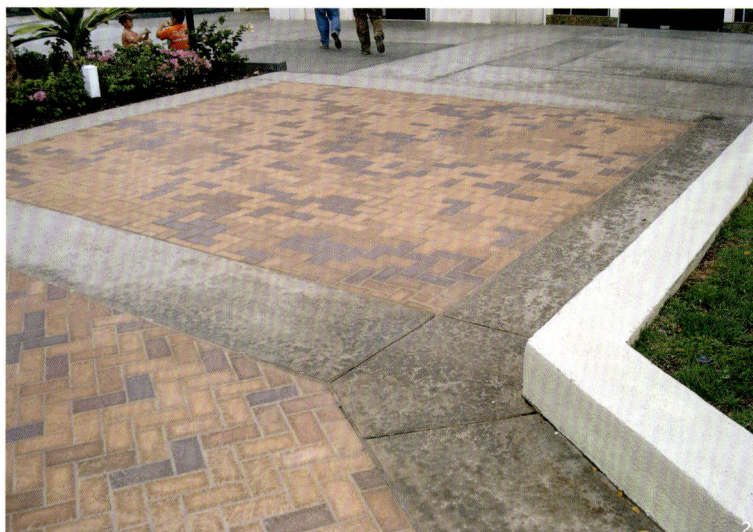

1、2　路面的铺装材料

3　女王购物中心美丽的景色吸引人们驻足观赏、留影

4　女王购物中心里静谧的花园

Queens' Market Place is a 12,500-square-meter mixed-use resort commercial center located within the heart of Waikoloa Beach Resort on the west side of the Island of Hawaii. The resort is a world-famous tourist destination that was carved out of a vast ancient lava flow from a nearby and still active volcano. In this unique and challenging place, a beautiful shopping complex and tropical garden paradise has been realized.

Working in conjunction with Hawaii architect Ted Garduque, FAIA, of Garduque Architects, the landscape architects at Helber Hastert & Fee were responsible for planting, irrigation, and lighting design as well as conceptual and aesthetic direction for hardscape and water features. The intention of the landscape design was to create a sequence of landscape spaces and destination points within the mall area

to provide a memorable and varied experience for visitors. The garden features include naturalistic ponds with fish and water plants, laminar shooting jets, a bubbling "volcano" feature, and a majestic granite fountain with pineapple finial. Landscape furniture and architectural elements include a focal clock sculpture, decorative fencing, and a large garden gazebo and open lawn that serves as a performance and event area. The use of colorful and fragrant plants and plants native to Hawaii was emphasized in the design.

Before construction of the project, there was no existing vegetation or other features on the site except for the black, course lava rock, which gave the area the semblance of the planet Mars. Large quantities of soil and plants were imported from elsewhere on the island to create the lush and beautiful gardens. This coastal area of Hawaii is known for its abundant year-round sunshine and warm weather. The configuration of the mall and strategic location of shade trees, including coconut palms, helps to protect the gardens from the dry tradewinds that blow across the adjacent open lava fields. An automatic irrigation system with full coverage of all planting areas is essential. The plant material used on the project represents over 100 different tropical plant species, some of which are found only in Hawaii.

The architecture and landscape design of the Queens' Market Place has a theme that relates to Hawaiian history, in particular the queens of the royal monarchy that ruled the islands in the 1800s. The design emulates an elegant style of living and romantic tropical garden aesthetic of that era. The design style also relate to a specific period of trade between Hawaii and China. Architectural features, such as a pagoda style roof on one building, stone clad walls, and the use of large ceramic pots for vines, relates to the cultural influence that China has had on Hawaii.

An interesting feature of the landscape design of Helber Hastert & Fee was the creative use of different paving materials in the mall, including quartzite flagstone, black slate flagstone, and multi-colored brick pavers, and stained concrete paving. Irregular joint patterns and the application of rock salt and dark stain to the concrete paving finish create a pleasant and almost naturalistic quality that relates to the native lava stone of the area.

The water features are an integral part of the landscape architecture and reinforce the tropical "paradise" atmosphere of the gardens. A "laminar" jet water feature is a particularly fun element in the garden and acts as a memorable focal

point. Long streams of water shoot up in a random sequence, appear suddenly out of beds of rock, and arc across pathways and through plantings to land in ceramic pots and into a pond. This feature is a great attention getter, and amazes and delights the surprised shoppers, particularly children.

A formal fountain with a carved stone pineapple top creates a welcoming greeting to the main entrance of the mall. The fountain is made of granite and was carved in China. The formal fountain is a nice contrast to the more naturalistic fountains found elsewhere on the project and helps to relate the landscape to the historic theme.

The conceptual design of the water features was by Helber Hastert & Fee, who communicated the design intent with a series of sketches, as well as plan view drawings. The installation of the water features, including the design of the pumping and filtration systems, was done by the water feature specialists Kai Pono Builders of Hawaii. Kai Pono Builders used a variety of materials and techniques to create beautiful pools and waterfalls that look like they are almost natural features on the site. A dramatic "lava stone" bridge across a Koi pond in the garden is a good example of the artistic capability of the contractor.

1 不规则的地砖给人以舒适自然的感觉
2 大钟成为花园中的景观焦点
3 女王购物中心草木茂盛，风景优雅
4 葡萄藤爬上了商铺的墙壁

生活方式购物中心 —— 北山购物中心

Lifestyle Shopping Center — North Hills Shopping

撰文：Scott Rykiel　　图片提供：J Brongh Schamp　Jacobs　Mahan Rykiel　Associates Inc.　　翻译：孙禹

Mahan Rykiel 公司与建筑师 Carter Burgess 通力合作，共同打造出一座生活方式购物中心——北山购物中心。设计借鉴了美国小镇购物街的模式，铺设了可举办大型活动的大片的公共绿地，并预留了就餐和橱窗购物的路边空间。目前，这种生活方式购物中心已成为商业区的一种流行趋势，它将新的生活方式融入到传统的大型购物中心之中，许多购物中心正向这种模式转变。

北山是罗利市（美国卡罗莱纳州北部）最早的封闭式购物中心之一，客户决定为这一古老的商业中心注入新的活力。经过与客户和建筑师的沟通，设计师在 125 453 m² 的场地上营造了一个开放的购物空间，为喜欢购物、休闲、看电影的人们提供了优美的环境。设计师在场地中央建了一块类似于中心广场的开放空间，四周的行人和车辆川流不息，这块空间可被灵活使用，既方便政府在此举办大型晚会又可用做市民的纳凉之地。该项目促进了零售业和娱乐业的繁荣发展，能够满足不同顾客的使用

需求。设计师还在街道下设计了双层停车场，以满足更多人对停车位的需求。

把主要街道建在停车场的上面给景观设计带来了巨大的挑战。设计师注重细节和高水平的景观维护使该项目有别于其他的购物中心，通过使用过渡性景观把主要街道与中心连接在一起，为人们营造了一个舒适的购物环境。该项目的成功已被公众认可，并带动了相邻商业区的发展，成为了一个名副其实的商业中心。

总平面图

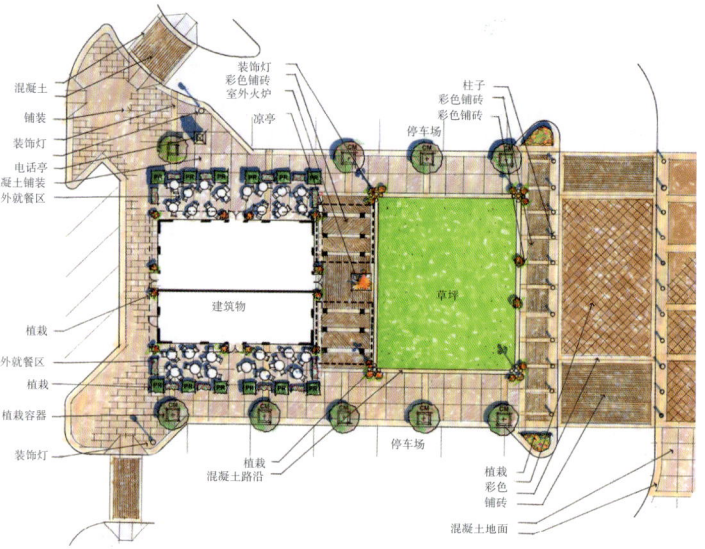

公共绿地的总平面图

Mahan Rykiel Associates worked with the architects, Carter Burgess, to create a new lifestyle center, similar to "Main Street" in small town America from an existing enclosed mall. There is a common green for events and curbside spaces for dining and window shopping. This lifestyle center is a popular trend in retail design, currently, breathing new life into traditional shopping malls. Many malls are being redeveloped in the model of a "Main Street" with pedestrian and vehicular traffic side by side and a town square type open space at the center. The town square serves as programmable flexible open space to be used by the local government and owner for public gatherings.

North Hills was one of the first enclosed shopping malls in the Raleigh area. Developer John Kane decided to breathe new life into this aging property. Working closely with the owner and architects, Mahan Rykiel created a small town setting for this 31-acre site. Located in the suburbs of North Carolina, the center provides enhanced venues for

1　公共绿地为音乐会、商品展览会和交易会提供了场所
2　临街停车场营造出购物街的氛围
3　咖啡店和冰淇淋店是人们夜间的活动中心

北山购物中心立面图

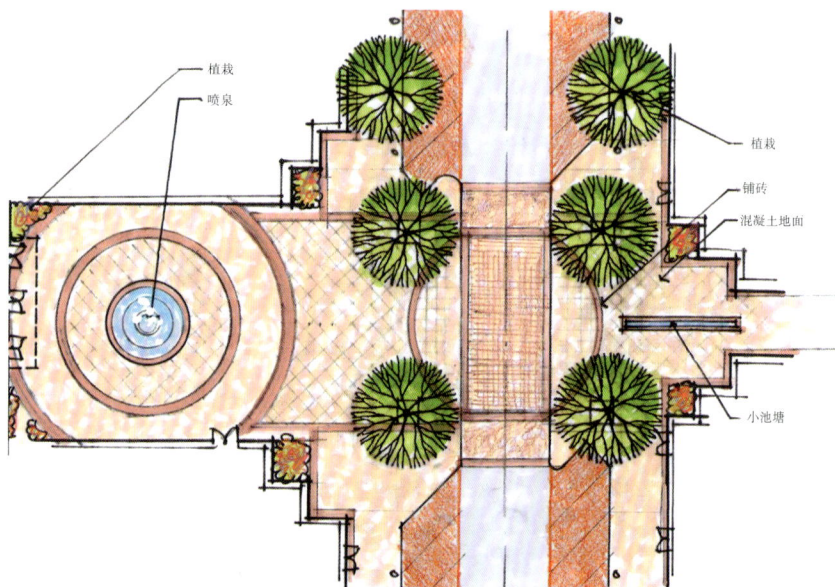

办公区入口广场的规划图

植栽

喷泉

植栽

铺砖

混凝土地面

小池塘

1　开放空间的设计使人们能在户外用餐和休息

2　大型的表演舞台

3　鸟瞰图

4　中央喷泉处的圣诞树

shoppers and movie goers alike. MRA was responsible for the streetscape design, plazas, and fountains and detailed planting for this complex. The project includes more than 770,000 square feet of retail and restaurants laid out around town square and shopping streets with on-street parking. The project also includes a cinema, a hotel, offices, and 300-unit residential building and luxury condominiums. The mixed use nature of the tenants guarantees a vibrant destination to support the retail and entertainment venues within the project. This increased program required more parking which could only be accommodated with structures; consequently two levels of parking exist below much of the Main Street.

Creating a main street over a parking deck, presented many design challenges for the street trees and landscape features. The attention to detail and the high level of horticulture maintenance set this project apart from most lifestyle centers. A main street/avenue design links the center using consistent materials that provide an authentic environment. North Hills has proven to be very successful and has been a catalyst for additional development on adjacent sites, making it truly a town center.

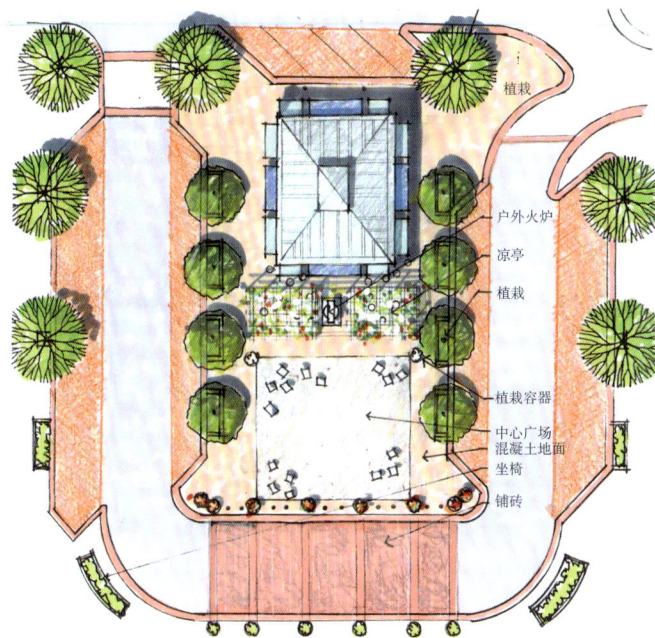

中心广场早期的规划图

戴维斯太平洋中心

Davies Pacific Center

撰文：Robyn Sweesy　　图片提供：Richard Quinn　　翻译：曹亮

1　广场上的小喷泉
2　俯瞰戴维斯太平洋中心广场

戴维斯太平洋中心位于夏威夷檀香山市的商业区，是一座拥有35年历史的高层写字楼和商品零售商厦。大厦占据了商业区的整条街道，在它的主入口处有一个500 ㎡的广场，面积虽然不大，但其景观设计为该商业区增添了一抹亮丽的色彩，为城市注入了一股新的活力。2003年12月，戴维斯太平洋有限责任公司购买了这个大厦，通过对其进行修复、重建从而实现地产增值。设计师认为原广场并没有发挥其应有的作用，应对其进行重点改造，但修建过程中不会拆除整个广场，而是要将现有的特色以一种新的方式展现出来。

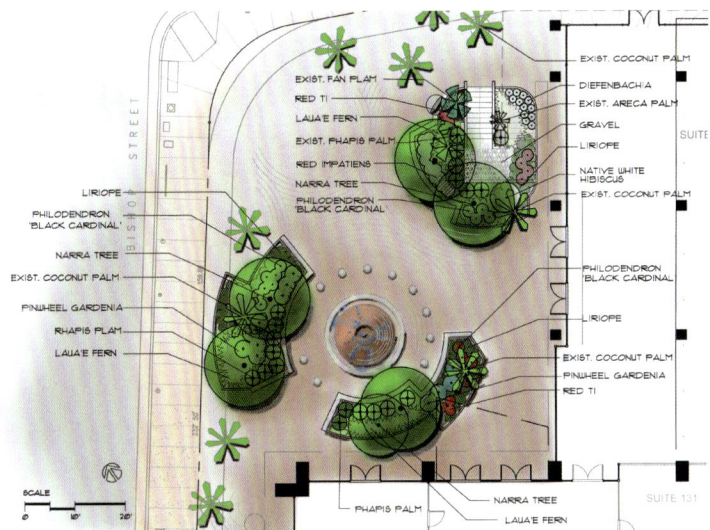

总平面图

最初，广场是按照现代园林的抽象风格设计的，然而建成之后的广场却门可罗雀、了无生气，只有少数行人匆匆穿过，既不能吸引人们驻足停留，也很难招揽人们进入商厦购物。广场上种植的椰树已经长成了参天大树，可是光秃秃的树干犹如电线杆，破坏了空间的美感，而那些设计精巧的地砖图案也未得到人们的欣赏与认可。靠近广场中心的小喷泉是一处景观焦点，给人以宁静舒缓的感觉，但单凭此处的喷泉根本无法吸引行人，因此改造势在必行。

2004 年，改造工作开始了。设计师理查德·奎因带领 Helber Hastert & Fee 公司的景观设计师们，开始对包括小广场在内的整个戴维斯太平洋中心的景观进行升级改造工作。

广场的改造就是要突出商厦的主入口，使其能够吸引人们的眼球，改善广场的微环境，让这片空间充满活力、生机盎然。要达到上述目标，需要整合广场中现有的人性化元素，同时弱化已有元素的随意性，保持广场及商业区外围的景观特色以及保留商厦的广告牌和橱窗。

为了更好地改进广场上的景观，设计公司提出了很多设计方案，如增加一个大型雕塑或新建一个喷泉，亦或重新设计一个新广场。最终，设计师决定将现有的喷泉作为商业区的一个重要景观元素，并保留该项目中同心圆图案的地砖设计，设置新的花坛和矮小的围墙以增强广场的动感。这些新增的特色景观体现了广场的人性化设计，拉近了人们与广场之间的距离。

混凝土筑成的小石凳、商厦的外立面、喷泉周围的弧形花坛以及广场的台阶彼此间相互映衬，石凳和围墙错落有致，这些细微的变化使这里的一切充满活力和动感。

新栽种的紫檀树为广场上的人们提供了阴凉，同时也为大厦遮挡太阳，使它们与整个广场完美地融为一体。此外，其他新的灌木品种和地被植物，如棕竹、栀子花、喜林芋等，以及夏威夷的本土植物，把广场装扮得五彩斑斓、富有生气，并营造出和谐统一的氛围。这些植物的品种和种植位置都经过设计师的精心设计，以此来保持广场内外景色的和谐。

夜幕降临，广场周围的路灯和建筑灯光会将广场照亮，因此广场内部只需在石凳和花坛的围墙上镶嵌低压灯泡，利用这些柔和的光线映衬出广场内的环形路线。

精湛的工艺、合理的材料搭配、和谐的外观使戴维斯太平洋中心焕然一新。简洁的线条清晰地勾勒出广场的轮廓，使其成为檀香山商业区鳞次栉比的高楼大厦中的亮点。

新广场整体结构的设计灵感来源于火山状的喷泉，无论是视觉效果还是亲身体验，广场都散发出无穷的魅力。广场上的景观相互呼应，构成了一幅绝佳的风景画。

Davies Pacific Center is a 35 year old high rise office and retail building in downtown Honolulu Hawaii. The building occupies an entire city block in the central business district and has a 500 square meter plaza that marks the building's main entry. Although relatively small, the plaza represents a good example of how landscape renovation can change the character and vitality of an existing urban space. In December 2003, Davies Pacific Center was purchased by Davies Pacific, LLC of the The Shidler Group, which is well known for buying existing properties and increasing their value through renovation and repositioning. The new owners looked carefully at the plaza and decided that the plaza was not working as it should. Rather then demolishing the entire plaza and starting with a fresh "canvas", a renovation design was developed to capitalize on existing features and make them work in a new and more unified way.

Although the original plaza area had been carefully designed at the time within the abstract style of the modern garden movement, the reality of the plaza was that it was an inhospitable and sterile transition zone that people moved through quickly. The plaza did not encourage people to stop and enjoy the space, nor did it present the building entry in an inviting manner. Coconut palms that had been planted in small round cut outs in the plaza had grown to great heights and their bare trunks were like scattered poles that interrupted the space. The brick paving pattern, although nicely designed, was a barren expanse. A small water feature near the center of the plaza gave some relief and an element of focus, but was not affective enough by itself to save the plaza.

The renovation process began in 2004. The landscape department of Helber Hastert & Fee, Planners, led by Richard Quinn, a Hawaii landscape architect, was selected to design the landscape improvements for the entire Davies Pacific Center block, including the plaza.

The objectives of the plaza renovation were to enhance the building's main entry, to improve the plaza microclimate, and "activate" the space. Essential to those goals was to integrate human-scale elements and mitigate the randomness of existing plaza elements; maintain views from the plaza to the downtown environs and views to building retail signage and windows; and obtain the best value while satisfying these objectives.

For the plaza landscape improvements, Helber Hastert & Fee explored many design alternatives, from adding a major sculpture piece or a new fountain to replace the old one, or to replace the entire plaza with a new design. The final decision was to celebrate the existing fountain as an important downtown feature and retain the concentric circles of brick paving that are used throughout the exterior areas of Davies Pacific Center block, while incorporating new walls and planter cut out in the paving to change the plaza dynamics. New features were designed to reinforce the plaza and the human-scale references.

Low seat walls—constructed of concrete to match the building exterior—and curbed planting beds arc around the fountain and the plaza stairs. Subtle variations in the heights and placement of seat walls and curbs make the elements lively and fluid.

New narra trees (Pterocarpus indicus) shade the plaza, function as a solar buffer for the building, and create a unified appearance and inviting entry to the plaza. New shrubs and ground cover—including rhapis palms, pinwheel gardenia, philodendron, ti, and several native Hawaiian plants—provide an important human-scale reference and a variety of colors and textures. Clusters of similar plant species throughout the plaza create a sense of balance and unity. The species and location of plant material were carefully selected to maintain views into and out of the plaza.Three existing large flagpoles in the plaza were determined to be out of scale with the size of the plaza and were removed.

Since a large amount of ambient light from street lamps and architectural lighting illuminates the plaza at night, the lighting design includes only low-voltage, seat wall-mounted niche lights installed to subtly accent circulation routes through the plaza.

Good workmanship, materials that match existing elements, and the unity of shapes, create a seamless connection between the 1971 plaza and the 2007 plaza. Clean lines and clearly delineated zones make the plaza an easy-to-navigate space in the dense urban fabric of downtown Honolulu.

With the overall structure of the new plaza taking its cue from the "volcano" fountain, the visual focus of this fountain is elevated to a new level of significance, and the unity of the plaza is visually and experientially appealing. It is now an eye-appealing panorama of green planting reverberating with the complementary color of the brick pavers. Although the plaza was significantly changed, the essential identity of the plaza was preserved and strengthened.

维多利亚花园

Victoria Gardens

撰文：SWA Group   图片提供：Tom Fox   翻译：孙禹

维多利亚花园在 2006 年获得了由城市土地研究所颁发的优秀奖，成为加利福尼亚州快速发展的杰出典范。此外，在为数不多的大型综合项目中，它被该机构评为具有代表性的项目之一。

SWA 在伦秋库卡蒙加市建成了一个全新的市中心区。该区域拥有 30 个街区，商店、办公楼和住宅楼沿传统的街道格网布局，彼此相邻，交通便捷。户外广场、民用设施以及宽阔美观的步行街使新的市中心备受欢迎，并给城市注入了新的活力。

该项目的设计将美国西部小镇中心的优秀传统与 21 世纪初期的社会要求和规划要求完美地结合在一起——零售店、办公楼、住宅楼、民用设施和文化场所都坐落在美丽而又古老的街道旁。维多利亚花园的景观框架以当地原有的街道和步行街为基础，城市广场、林阴大道和公园点缀其中。

维多利亚花园是美国第二大露天购物中心，仅次于科罗拉多州丹佛市的 Stapleton，二者均由 Forest 城市开发商建造。

位于加利福尼亚州"内陆帝国"的伦秋库卡蒙加市期望在未来的 15 年内人口增长约 400 万。该市有着悠久的殖民地历史，从最初的库卡蒙加村到西班牙的殖民地，随后又成为墨西哥的大牧场，最终于 19 世纪中期成为美国的领土。6 条主要贸易路线经过这里，包括最便捷的路线 EI Camino Real、连接科罗拉多河及以西地区的重要贸易路线 Santa Inez 以及二战后期在西部大移民中起到重要作用的 66 号公路。

该项目在硬质景观和种植材料方面充分展现了当地的历史渊源，成为了一座独一无二的购物中心。花园的入口是一个标准的枣椰树凹槽，确定了该项目将提供给人们丰富景观体验的基调。格网布局的棕榈树提醒人们这里曾是高产的果园和田地，同时高大的树木也成为高速公路与市中心之间的屏障。此外，华盛顿棕被用做环绕整个 647 520 ㎡ 项目场地的屏障和标志。

1

1　线形喷泉
2　傍晚时分的维多利亚花园

总体规划图

Washingtonia robusta & Pyrus calleryana "Bradford"
Washingtonia robusta & Magnolia grandiflora "Samuel Sommer"
Washingtonia robusta & Rhus lancea
Eucalyptus citriodora
Schinus molle
Pyrus calleryana "Redspire"
Ulmus parvifolia
Phoenix dactylifera
Platanus acerifolia
Washingtonia robusta
Bauhinia purpurea
Cassia leptophylla
Chitalpa tashkentensis
Jacaranda mimosifolia
Koelreuteria bipinnata
Koelreuteria paniculata
Lagerstroemia spp.
Laurus nomilis
Magnolia grandiflora
Olea europaea
Phoenix canariensis
Prunus spp
Quercus spp
Robinia ambigua "Idahoensis"

设计师在项目周围设置了充足的停车位，市中心的交通压力得到了很大程度的缓解。为了避免上百辆车聚集一处而产生突兀感，该项目利用一排排的花架来营造出果园般的场地效果，而植栽也为场地营造出多样的柔美效果。南北主街道两旁种植了具有浓郁乡村气息的悬铃木，其他街道旁也排列种植着加州胡椒树、桉树、非洲漆树、Aristocrat Pear 等，参天的古树则传达出该项目建在已有景观之上的信息。不同种类的遮阴树提高了花园的质量，众多的公园、广场和林阴大道为人们提供了更多适合散步的空间。维多利亚花园里的大多数树木都种植在原有的本土灌木丛和草地里，重新营造出当地在殖民地初期的典型景观，并以此突出花园的历史特征。

水景是设计中的主要元素，设计师在每个重要的位置上都设计了各式各样的喷泉。小型水景墙使人们置身于优雅的环境中，耳畔回荡着延绵舒缓的水声；稍大的石灰岩庭院喷泉周围是带有彩色装饰的可坐矮墙，为游人提供了一个凉爽、宁静的聚会场所；线形喷泉用光滑的石头围边，喷水时溅在石头上的水能产生一种特别的声音效果，如同夜晚时分墨西哥小镇广场上的灯光协奏曲。

无论何时，维多利亚的广场和林阴大道都能为伦秋库卡蒙加市注入一股新的活力。

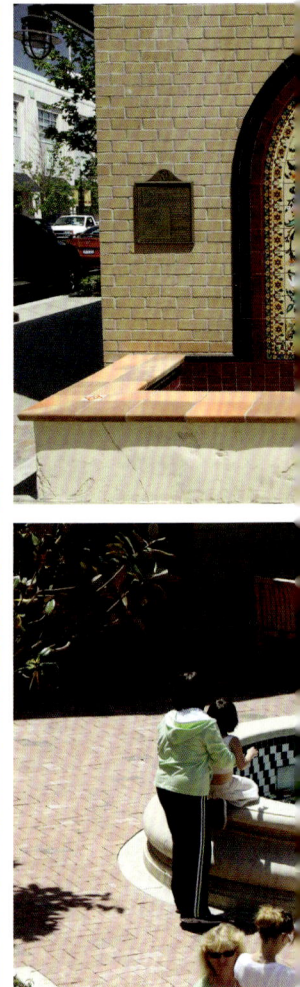

Victoria Gardens received a 2006 Award of Excellence from the Urban Land Institute as a leading example of smart growth for California. In addition to the award, Victoria Gardens was also selected by the ULI as one of a handful of projects to serve as case study examples in a publication on mixed-use projects.

SWA created a new town center that occupies 30 square blocks in the city of Rancho Cucamonga. Organized along a traditional street grid, shops, offices and residences are all located within easy walking distance of each other. Outdoor plazas, civic uses, and wide, landscaped sidewalks have made this new downtown core a popular destination and brought renewal to the urban center. Once the Master Plan was approved by the City of Rancho Cucamonga, a number of leading retail architectural firms were invited to design individual buildings within the project.

This SWA-designed downtown combines the best traditions of western American town centers with the social and planning demands of the early 21st century. Retail, office, residential, civic and cultural uses are placed within the landscaped urban experience of a traditional main street environment. The landscape framework for Victoria Gardens is a grid of local streets and sidewalks with a town square, plazas, paseos, and parks distributed throughout the downtown district. SWA provided full landscape architectural planning, design and implementation services.

Victoria Gardens is the second largest open-air shopping center in the United States behind Stapleton in Denver, CO, which is being built by the same developer, Forest City Development.

The city of Rancho Cucamonga, which is located in the Inland Empire in California, is expected to grow by about 4 million people in the next 15 years. The area has a rich history of settlement from the original native Kucamongan villages to the Spanish missions and later the Mexican rancho settlements and finally to annexation and settlement as a part of the U.S. in the mid-1800s. Six major trade routes passed through the area including the El Camino Real (connecting the missions by a days travel), Santa Inez Trail (major trade route connecting the Colorado River to the settlements out West), and Route 66 (a major cross-country highway route that enabled mass migration westward during the dustbowl and post WWII era).

Victoria Gardens has become a unique shopping center by tapping into the region's historical roots in both the hardscape

design elements, and in the planting materials and design. The entrance to Victoria Gardens, a formal Date Palm groove, sets the tone for a rich, textural landscape experience. The formal grid arrangement of the palms recalls the region's agrarian roots, which was once prolific with orchards and fields. The mature trees also help to screen the city center from the freeway system nearby. In addition, Mexican Fan Palms are used as screen and as a hallmark along the entire perimeter of the 160-acre site.

The downtown area is spared the impact of vehicular traffic the use of cars within the urban center by proving ample parking on the outer rim of the project. Convenient to the core, the parking areas are designed to limit the obtrusive visual impact of hundreds of parked cars by using linear planters of Bradford Pears to create an orchard-like effect. The variety

and softening effect of the planting is seen throughout the site. Both North and South Main Streets are lined with London Plane trees, which have a rustic quality marking the passing of seasons with their fall color. Other arterial streets are lined with California Peppers, Eucalyptus, African Sumac, and Aristocrat Pear. Large heritage trees are planted throughout the site in such a way as to create the impression that this project was built within and around an existing landscape. Several different species of accent shade trees are featured to enhance the garden quality of the project while adding a pedestrian scale to the numerous, parks, plazas, and paseos. A majority of the trees throughout Victoria Gardens are planted in beds of native shrubs and groundcovers to further enhance the garden heritage of the project by reestablishing a landscape typical to the early settlement era of the region.

Water is a major element in the site design. There are a variety of types of fountains that are set at every gathering point. Small wall fountains enhance the environment both as visual inspiration and in providing a constant, soothing background of sound. The larger limestone courtyard fountains each have seat-walls with colorful tiled interiors providing visitors with a cooling restful gathering spot. The linear runnel fountain is lined with smooth stones, and as in a creek bed, the water splashes over the stones for a special sound environment when the fountain is running. It reflects the strings of lights at night just as on any Mexican town square.

Night and day the plazas and paseos of Victoria Crossing have become a hub of new activity in the city of Rancho Cucamonga, California.

Chía商业中心公园

Park at Chía's Commercial Center

撰文：Jimena Martignoni　　图片提供：Olga Lucia Jordan　Claudia Misteli　Martha Fajardo　　翻译：李博

20世纪90年代末，哥伦比亚政府要求每座人口超过50 000人的城市都要制定土地使用规划，使人们能够合理地进行土地规划并轻松地应对自然和社会的变化。基于这种社会背景，Chía城制定并实施了一系列土地规划，以扩大土地的使用面积。Chía城位于波哥大市北部20km处，是许多市民向往的安静居所。因此，在过去的15年内，Chía城内新建了许多住宅区与商业区，而作为整体规划的一部分，Chía商业中心诞生了。

该项目原址是一个农庄，所有者将其作为不动产加以发展和利用。规划共分为三个阶段：第一阶段建造商业中心；第二阶段建造了128栋公寓；第三阶段，又建造了125栋公寓，并对商业中心进行了扩建，主要内容是建造公园与绿地。

哥伦比亚著名景观设计师、前IFLA主席Martha Fajardo受邀参与了此次规划，该项目的方案于2006年通过，施工历时9个月，于2007年3月对公众开放，成为了一处充满欢乐的天堂。Martha Fajardo说："打造一处娱乐之所是此次设计的首要目的，孩子们可以在这里尽情嬉戏，大人们也会因此而快乐。因此，这是一处流淌着欢乐与笑声的天堂。"

整个规划分为三个阶段：调查与测评、概念规划、细节设计。第一阶段，设计师对场地及原有植物进行了调查，确定最佳发展方案；第二阶段，设计师制定了一个完整的规划来表达"欢乐"这一理念；第三阶段决定设计与建造方面的细节问题。

尽管占地25 000m²的公园是构成"欢乐"理念的主要元素，但设计师还是在场地中营造了休闲轻松的氛围，田园式布局的中央湖泊尤其体现出这一点，

总平面图

水景和伸向游乐区的两侧有植栽的小路则营造出安静的氛围。设计师尤其注重安静与活力二者之间的平衡，使其形成鲜明的对比，并努力为附近的住宅区提供私密而静谧的环境。设计师还设计了一片林木，作为住宅区和商业区之间的过渡带。

孩子们的活动区域包括特制的形式新颖的游乐场，可供5岁~12岁或刚学会走路的孩子使用。设计师为游乐场选用的色彩与场地内的植栽保持协调，如蓝色的百子莲与浅蓝色的攀爬设施相呼应，火炬花与小孩子钟爱的浅红色的攀爬网相呼应。

植栽规划包括两个主题：树木结构、水生植物与禾本科植物。在树木结构方面设计师选用能够丰富景观色彩的树木，如枫香属、栎属和巴豆属植物等。在设计扩建区域的自然空间时，设计师移植了一部分场地原有的树木，并保留了部分不会影响到水景规划的树木。如纸莎草、灯心草、日本鸢尾等水生植物被种植在湖边步道上，形成不同的植物群，营造出多样的景观。为了体现湖畔的自然景观，此处区域主要种植了禾本科植物，如狼尾草与蒲苇，成为一处理想的野鸭栖息地。

场地内原有的湖泊历经多次洪水后，雨季曾存在水位线大幅度上涨的危险。设计师对此采取的一项重要措施就是设计一个可持续水体，以控制洪水泛滥。设计师对土地进行了重新设计，将湖泊面积加倍，现在约为3994平方米，并设计了一个闭路循环排水系统，其中包括一系列沟渠和径流收集设施，将水引至湖中。

多余的湖水由水泵引至距场地约50米处的波哥大河中。设计师沿较高处的道路设计了一系列沟渠用来储存暴雨时的雨水。沟渠储存的雨水将沿坡流下，到达场地后方的线型排水系统，最后流入湖中。

　　一座木桥架于水面之上，还有一些较小的瞭望台和甲板与入口平台相连，在水面上形成了一个阳台。Chía公园中还建造了两个咖啡馆，一个沿湖而建，另一个位于儿童游乐区内。两处都建有可伸缩屋顶，形成了半露天空间，可供人们在此休憩、聊天。人们可凭湖欣赏美景，或在咖啡馆中小坐，或在游乐场中散步，或进行各种休闲活动。

　　每当夜幕降临，照明就成了商业中心的主要景观，吸引游客驻足于此。这里经常举办钢琴演奏会、歌剧演出以及各种家庭文化活动，人们在此流连忘返。照明包括四种类型：人行道照明（短柱形状的照明设施）、滨水木栈道照明（LED灯光）、植栽周围的照明（可勾勒出植物的轮廓）、建筑物上的照明（突显小桥、建筑物以及半露天区域）。

　　项目中的绿色开放区域成为吸引游客及潜在投资者的主要场所。因此，在当代的多用途开发项目中应多设计此类景观，为人们营造愉悦的景观环境。

At the end of the 1990s, every Colombian city which has more than 50,000 inhabitants was compelled by the national government to elaborate a Land Use Plan or Plan de Ordenamiento Territorial (the so-called POT) in order to responsibly plan and respond to all physical and social changes and needs within their own areas. Under these conditions, the town of Chía had to elaborate and put into action some plans for its expansion. Located 20 km to the north of Bogotá, Colombia's capital city, this town was the Mecca of numerous Bogota's citizens who sought for quieter yet not distant settlements. For this reason, new residential communities have been built and new commercial areas have been implemented in the last fifteen years; as part of this process and as a response to the necessity of expansion of the downtown, Chia's Commercial Centre (Centro Comercial Chía) was born.

The site itself was a "finca" (farm) whose owner divided and developed for real-estate purposes. The first phase was the construction of the commercial centre; a second phase finished 128 residences and a third one is in the final process of building 125 residences more, plus the expansion of the commercial area. The construction of a park or green areas for the complex was one of the main conditions for this expansion's plan.

Initially, the owners of the site called a firm of Canadian consultants called Commercial Design Corp and STOA, a local architecture firm; the first presented a conceptual Master Plan and the last would be in charge of the implementation. However, when the project was presented to the Resident's Committee or Junta de Vecinos of the complex, many items were either questioned or rejected and a new proposal had to be made in order to reach an

agreement between the parts.

Colombian landscape architect Martha Fajardo was part of this committee and was then invited to present an alternative project. Proposed and accepted in 2006, this landscape architecture project was definitely commissioned and the construction was finished in nine months, opening to the public in March 2007.

Martha Fajardo, who obtained her Master degree on Landscape Design in Sheffield University, England, was founder of the Colombian Society of Landscape Architects in 1983 and elected president of IFLA (International Federation of Landscape Architects) for two consecutive terms; her first presidency commencing in September 2002 and her second being completed at the end of 2006. Her office, called Grupo Verde Ltd, is currently based in Chía and is shared with urban planner and international consultant

Noboru Kawashima. Their project was specially created as a "place for fun and playing". According to Martha Fajardo's words: "Playing is the action which we decided to take as a kick-off idea for this place…The kind of playing that children want and generate with their actions…, but the entire project is thought of as a place for fun. If kids are happy, parents are happy".

Split into three different phases that respond to "Inventory and Evaluation" "Conceptual Plan" and "Details", the overall plan was completed and presented to the investors and the complex' board in timely manner. With the first phase, they carried out an extensive survey of the site including lists of existing trees and plants and also established best development options; with the second, they elaborated an in-depth plan whose catalyst concept was that of "playing" and with the last one determined design and construction details.

Although the key concept of 2.5 hectare-park is that of playing, the site also offers a relaxation-defined profile, particularly appreciated in the bucolic layout of the central lake. The park presents tranquil spots framed by the water and planted paths which complement the areas for playgrounds and spots for kids' gathering; this contrast between contemplative and active areas was carefully studied in order to preserve the privacy and serenity of the adjacent residential lots. For this reason, the quieter areas were laid out nearby and those characterized as noisier occupy more distant spots, which are actually adjacent to a small existing wood-like formation that act as a natural buffer.

The kids' areas incorporated some innovative playgrounds which were specially brought for the site and are differentiated in two groups, one for 5 to 12 year old kids and another for toddlers. The colours of these play-systems were intentionally matched with the planting plan: masses of blue agapanthus frame the bluish structures where kids pretend to scale and borders of eye-catching red-hot poker (Kniphofia uvaria) accompany the reddish climbing meshes that little ones happily conquer.

The planting plan is completed with two other thematic compositions: "tree structures" and "aquatic plants and grasses". The first includes all trees species that were added

primarily to incorporate color. Liquidambar or sweetgum tree, Quercus humboldtii or South American oak and Croton funckianus are some of the most interesting examples. In order to take advantage of a large group of trees that existed in the site, the designers decided to use most of them to generate new clumps in the newly designed natural areas, around a central remodelled lake. Many were therefore transplanted and a few others were left in their original position as long as they didn't interrupt the new layout of the water surface. The aquatic plants, such as Cyperus papyrus, Juncus sp (rush) and Iris japonica (Japanese iris), frame the boardwalks positioned around the lake and appear as isolated groups, thus creating particular points of interest. In order to emphasize the naturalistic-look of the water's edges, the planting plan for these areas is mainly based in the use of grasses. Pennisetum and Cortaderia selloiana or Pampas grass visually lead the scene and create the ideal refuges for the funny families of ducks that walk around during the day.

The small lake that existed in the site, previous to the park construction, had suffered some floods and risky raising of the water's levels during rainy seasons. As a result, one of the key measures of the designers was the creation of a sustainable water body that could act as a natural flood-control element. They remodeled the land and doubled the size of the lake which now covers an area of 43,000 ft^2 and, in addition, they designed a closed-circuit drainage system with a series of runnels and runoff collectors that channel the water towards the lake. "In the past," says Martha Fajardo " when the percentage of construction in this area was much less, storm water was naturally absorbed by the soil and, sometimes, bombed and channeled to the Bogotá River… But now we need to help this process for it to be successfully completed."

The lake's exceeding water, which is now regulated by a pump that has a predetermined minimum and maximum water level, is pumped out towards the Bogotá River, something like 55 yards away from the site. On the other hand, storm water is retained in a series of runnels that extend along the site's upper paths. These paths were laid out on top of the green slopes artificially built with the material removed from the lake's remodeling; therefore they visually dominate the lake's waters from above. The water collected in the runnels is then channeled by gravity down through the slopes and towards a linear drainage which runs the full extent of the rear side of the site; from here, is the finally channeled to the lake.

A wooden bridge crosses over the aquatic surface and some smaller more intimate lookouts and decks, connected with the entry paved terraces, generate balconies over the water. Two café areas are also included in Parque Chía as part of its amenities, one by the lake and the other serving the kids' areas, both of which are organized underneath tensioned fabric roofs. In this manner, these spots become semi-open terraces that preserve the close contact of people, sitting and relaxing here, with the different spaces of the park. The double possibility of leisure and fun is something that visitors evidently enjoy and treasure, either by the lake, in a coffee-break, or at the many playgrounds this park offers.

In the night-time, lighting becomes a mayor attraction for the Commercial Centre's visitors. Piano concerts, opera sessions and family cultural activities are offered monthly and, in an everyday basis, people can enjoy the place until late in the evening. The lighting plan covers four different general topics: pedestrian paths (bollards that mark and illuminate the different footpaths), boardwalks (LED lamps), plant uplighting and dramatic and architectural lighting (to silhouette plants and objects and to create dramatic effects on bridges, buildings and semi-roofed areas).

In this commercial project, the incorporation of green open areas became a key "market hook" for visitors, possible shop owners and investors. Hence, what remains essential is that contemporary mixed-use developments should rely on these kinds of thoughtful landscape plans whose ultimate goal is people's pleasure.

办公区

墨西哥城联邦区科技园

Tecnoparque, Mexico DF

撰文：Jimena Martignoni　　图片提供：Francisco Gómez Sosa　　翻译：沈翀

该项目位于墨西哥城西北部阿茨卡波察尔科区的高科技园区内，该区域主要包括房价较低的住区、众多科研院所和一所大学，项目规划的目的旨在改变原来以轻、重工业为主的经济模式，并降低城市人口失业率。

项目占地约 15 万平方米，原本是一家钢铁厂，在 20 世纪 90 年代初被政府收回。2000 年，这里被私人开发商购买，用于建造办公大楼，政府同意了重新开发该地区的计划，但要求土地开发商长期提供第三产业就业机会，并尽量节约用水。

园区内包括了 6 栋平均占地面积为 6500 m² 的办公楼、可容纳 3500 辆机动车的停车场以及小型购物中心和员工餐厅。设计师在开放空间内建造了三座人工湖、花园、小路和停车场，并在每栋办公楼前的庭院内栽种葱郁的灌木，安置了设计独特的户外设施，营造出和谐的办公环境。场地内还设有服务中心，并对外开放。该项目现已成为一处市民休闲娱乐的城市空间。

水处理系统

在墨西哥城，对水资源的利用已经成为了十分重要的环保议题。因此，在该项目的整体景观规划及建设过程中，水资源的利用与管理已成为一个重要因素。水景的设计，不仅仅是景观设计中重要的视觉元素，还应体现其在可持续发展过程中应发挥的重要性。实际上，在墨西哥城的其他地区乃至全球范围内的新建项目中都应考虑到这个因素。

该项目主要包括了两种水处理系统：污水处理与

1

污水处理系统

① 洗手间
② 污水处理器
③ 蓄水池
④ 浇灌绿地

雨水收集系统

① 雨水收集
② 蓄水池
③ 深井

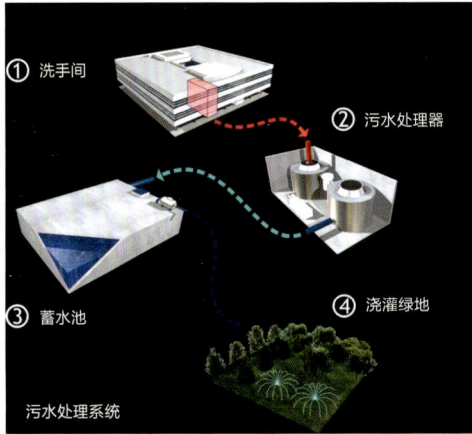

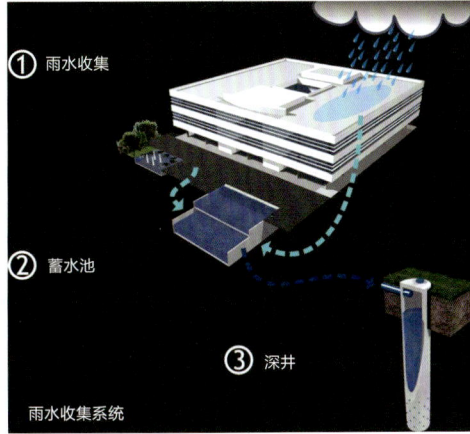

1 园区水景全貌

总平面图

循环系统、雨水收集系统，旨在最大限度地对园区内的水资源进行循环利用。污水处理与循环系统主要是对卫生间用水的处理，将过滤后的水用于花园和绿地的浇灌；雨水收集系统则主要收集屋顶及广场的雨水，利用火山岩将其过滤后可以浇灌树木和草坪，多余的雨水被直接排入城市污水系统。该雨水收集系统的创新点在于其能够将水回灌到地下蓄水层中，雨水首先被积蓄在一些大型的地下蓄水池中，然后将其排入到80m的深井中（以便深入地下岩层），过滤后的雨水从井底的孔洞中流出，地下蓄水层就这样被反复回灌。

景观设计

设计师将办公楼和花园交替设置，形成了由100m×100m的正方形构成的棋盘式的布局。该布局会使人联想到最初的西班牙城市布局，创造出简约而便捷的空间。独具匠心的设计使得园区内部的结构清晰明了，形成了便捷的步行线路，使行人可以在办公楼的柱廊旁与屋檐下漫步。这种内部结构与几何形步

道构成了办公楼的格局，设计师还在园区内设计了各具特色的花园和人工湖，营造出和谐的整体氛围。

市政园位于园区入口处，于 2005 年完工，主要采用了几何图形的设计理念。水景的面积约占市政园总面积的一半，是景观设计的重要组成部分。一条狭窄的步行桥建于水面之上，穿过草坪直达与环形步道相连的小路。水景区内对角线的两端是设计中较为突出的亮点：一处是如同在水面上漂浮的咖啡馆及其顶层的露天平台，功能多样，并设有桌椅、凉亭、快餐亭，观景角度极佳；而另一处则种满了纸莎草，犹如一个

1　炫彩夺目的美人蕉
2　工作之余，人们可以在园区内的咖啡馆放松身心
3　水雾喷泉
4　混凝土路面与绿地相间的道路铺装

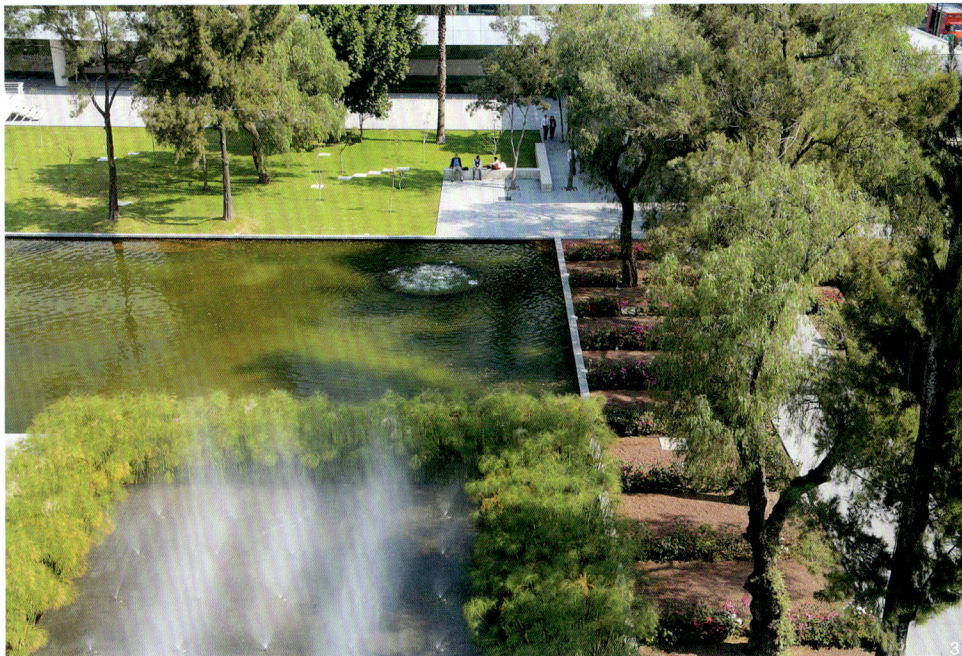

绿色方块，其间环绕着一组水雾喷泉，飞雾弥漫，水草丛生，使人沉浸于自然中，给人以视觉上的完美享受。硬景与绿地，人与自然和谐共生，“你站在桥上看风景，看风景的人在楼上看你……”

在咖啡馆旁摆放了几张坐椅供人们休憩。这些木质坐椅随意摆放，或临水放置，或沿路排列，或两两相对。市政园中还设有一座金属雕塑，形如火炬，高约22.86m。该雕塑矗立于园区内，高大醒目，因此成为了这里的焦点。

中央花园与市政园在视觉和空间上的关联体现了项目整体的设计理念——布局简易、结构清晰。网格形的步道将两处花园连接起来，而花园的有机布局更突出了这种简易、清晰的设计理念。中央花园位于园区内的中心位置，具有净化空气的作用，漫步其间十分惬意。

设计师还在中央花园内建造了一处不规则形状的人工湖，湖边种满了当地的草本植物。大簇的美人蕉炫彩夺目，给人以完美的视觉享受。园区内的植栽面积广阔，大片柔软的纸莎草与小径周围的草坪形成对比。

开放空间还包括一处停车场，周围种植的树木也体现了整体的布局方式。道路铺装采用混凝土与绿地相间的设计，以求最大限度地收集路面上的雨水。

周边环境设施

项目规划的另一个重点是在连接园区与周边交通线路和地铁的同时能保障安全性和舒适度。此处规划包括对不利于城市环境美观的周边设施进行设计及重建，诸如毗邻的地下通道。当人们从公交车站穿过地下通道，眼前呈现的便是干净整洁的景观区，走在混凝土与绿地相间的步道上，一堵弧形的金属景墙更使人们远离尘嚣。

在注重环境保护的墨西哥城，科技园已成为一处环境友好型的大型商务区。积极的、持久的概念化设计将该地成功地打造成一处绿色园地。人们可以在这里休息放松，并以更加愉悦的心情去面对工作。

Tecnoparque is a high-technology office campus located in the Azcapotzalco District, in the northwestern area of Mexico City, whose first construction phase was completed in 2005 and a second one in 2007. This area is currently characterized by the incorporation of low cost housing, technological institutes and a major university campus, all of which is part of a larger local plan whose objective was the eradication of a heavy and light industry-based economy and, especially, the decrease of mass departure and consequent loss of job posts.

The site itself, of approximately 15 hectares, was a steel production-related plant that was closed down and acquired by the City government in the early 1990s. In 2000, the land was sold to a group of private developers specialized in office buildings who negotiated the redevelopment of the site with the City; the project was approved with the condition to create permanent jobs in the tertiary sector and to use limited amounts of water from the city's lines.

The campus project proposes six office buildings—each one of 6500m^2—parking space for 3500 cars, a small shopping area and restaurants for employees; the common open spaces include three lakes with their gardens and pedestrian ways, on-grade parking and, additionally, the design of the central yards of each one of the buildings with harmonized masses of shrubs and specially designed outdoor furniture. At present, four office buildings and two of the lakes and gardens are completely finished; in addition, the site recently incorporated a service center which is

1　不规则形状的人工湖
2　植栽的设置增加了水景的自然魅力

opened for the site's users as well as to public at large, thus becoming a new people-attracting urban piece.

Because water supply and efficient consumption is a top environmental issue in Mexico City, this project's overall water management turned into one of the defining items of the landscape architecture Master Plan and site construction. Water not only appears as a main visual element of this office campus but as a highly relevant functional one that is coherent with the sustainable approach with which all new developments have to be thought out in this city, and probably should in any other of the planet.

In this case, there are two primary water management systems whose main objective is to maximize the reutilization of water within the site: the "black waters treatment and reuse system" and the "storm water collecting system".

The first one is based on the treatment of the restrooms water on site. The water is stored in pools and fountains placed underground and then, once recycled, serves to the irrigation of gardens and green areas. The second one is based on rain water collection from roofs of buildings and plazas. In parking areas, rain water is retained and filtered with lava rock to irrigate trees and grass pavers and excess water from roads is sent directly to city drainage.

But the water management plan here goes far and beyond and also presents an innovative system which conducts water back to the city aquifer. The rain water is first stored in some large underground retention cells and then sent to deep wells of approximately 80 meters long (thus reaching the deepest layers of ground) and whose bottom section is perforated to allow water to come out. In this manner, the city aquifer is constantly recharged.

The landscape layout is intimately connected with the architectural one. Buildings and gardens alternate, shaping a chest-like grid whose key module is a 100 × 100 meter figure; this specific size and grid arrangement reminds of the original Spanish city settlements and, in addition, creates a simple, easy-to-walk space.

This last situation was an explicit decision of the designers who wanted to allow an easy direction-finding within the campus. The grid pattern establishes clear pedestrian axis and allows for users to walk under the porticoes and overhangs of buildings. The identity and sense of place achieved with this model and the geometrical footway, which frames the buildings, is completed with the different composition of every one of the three gardens and lakes. One is the "civic" and entry garden, another is the "central" and natural garden and the third is the "still" garden; each garden providing eating facilities, seating areas and terraces and therefore creating opportunities to rest, talk, and meet other people.

The civic garden was the first one to be completed (2005) because it's the closest to the main vehicular and pedestrian entrances and consequently has a more institutional image. Strongly defined with a geometrical design approach, this garden is perceived as a very inviting, open area where the presence of water is also major.

The water surface occupies more than half of the garden area and the rest is a lawn-covered plane only interrupted by a couple of pines clusters. A narrow pedestrian path crosses the entire aquatic surface, as if floating over it, and then crosses over the green one to finally reach the paved paths that connect with the pedestrian circuit.

Two of the corners of the aquatic area, diagonally opposite to each other, create two different focal points: a functional one which is shaped as a floating semi-roofed deck or café terrace and a visual one modeled as a green square-like figure, built of papyrus, which frames a set of spray jets. The first one becomes a people-gathering space, with tables and chairs served by a kiosk and fast-food booth and with relaxing vistas towards the water, while the second turns into a nature-related element signed by the permanent ascending mist and the organic soft lines of the aquatic plants: a fine balance between hardscape and greenery or man and nature

1　连接园区与周边交通线路的步道
2　绿茵环绕的步行道
3　人工湖旁边的蜿蜒小路
4　随意摆放的坐椅可供人们休憩

and, from a different perspective, between a watching-place and a place to be watched.

At one of the sides of the water plane, next to the floating deck, a group of irregularly placed benches provide another spot for relaxation. Ergonomically designed as a one piece object, these wooden benches are randomly positioned to generate different social situations: some face the water, others face the pathway and others face one another.

In this garden it's also placed a metallic sculpture modeled as an inverted cone, approximately 75 feet-high, which stands as the site's icon. The verticality of this element deeply contrasts with the predominance of horizontal lines that delineate every one of the buildings and, accordingly, helps picturing this piece as a focal and reference point.

The second garden (2007) can be made out from the first one, assuring the visual and spatial connection that responds to the aforementioned leading concept of easy place-identification and direction-finding; the general grid-patterned pathway, on the other hand, connects also the two patches. However, this garden has a decidedly organic layout which validates the idea of nature with which it was created. Its central location within the campus also reinforces the relevance of a natural area that acts as a "decontaminating" green lung and which users enjoy by walking around or sitting at the semi-roofed café located here.

This central garden presents an irregular shaped lake whose curved edges are framed by widespread groups of native herbaceous plants. Acting as a main focal

1　园区夜景
2　园区内设置了供人们放松的休息区
3　漂浮在水面上的咖啡馆

arrangement, a large group of orange, yellow and red Cannas indicas (or Indian shot) produce a fantastic eye-catching effect and provide the perfect spot for visual recreation and relaxation. The completing planting is primarily based on other extensive punctual groups, such as the soft stylized papyrus, which contrast with other horizontal lawn planes that follow the curved paths of this place.

The rest of the open spaces are occupied by the parking areas, all of which are planted with trees whose planting pattern echoes that of the site's general grid layout. In order to increase the rainwater absorbing surfaces, parts of the pavement are built as a net that alternates concrete and lawn. All in all, these areas complete the landscape plan and the green network that supports the complex.

Another relevant concern of the Master Plan was to provide comfort and safety through connections with the transit lines and metro subway system, including the effort to design and transform surrounding detrimental urban environments such as the adjacent underpass bridge. The place to which pedestrians arrive at, after crossing this bridge that connects with public transportation stops on the adjoining main road, is another neatly landscaped area that alternate paved and planted strips and which is visually separated from the street with a curvilinear metallic wall.

Tecnoparque becomes an environmentally-friendly large business area in a city where environmental issues are on top of the list. This positive, hopefully lasting, conceptual design aspect of the project adds up to the knowledgeable decision of providing the site with highly attractive green areas where working people can find respite, recreation and, let's face it, a good reason—other than responsibility—to go back to work the next day.

康菲石油公司总部

ConocoPhillips Company

撰文：The Office of James Burnett   图片提供：Hester+Hardaway   翻译：张咏梅

办公楼

办公楼

足球场

庭院

庭院

停车场

健身中心

庭院

办公楼

Employee Entry

接待中心

Main S

办公楼

办公楼

庭院

休闲区

员工入口

访客入口

办公楼

总平面图

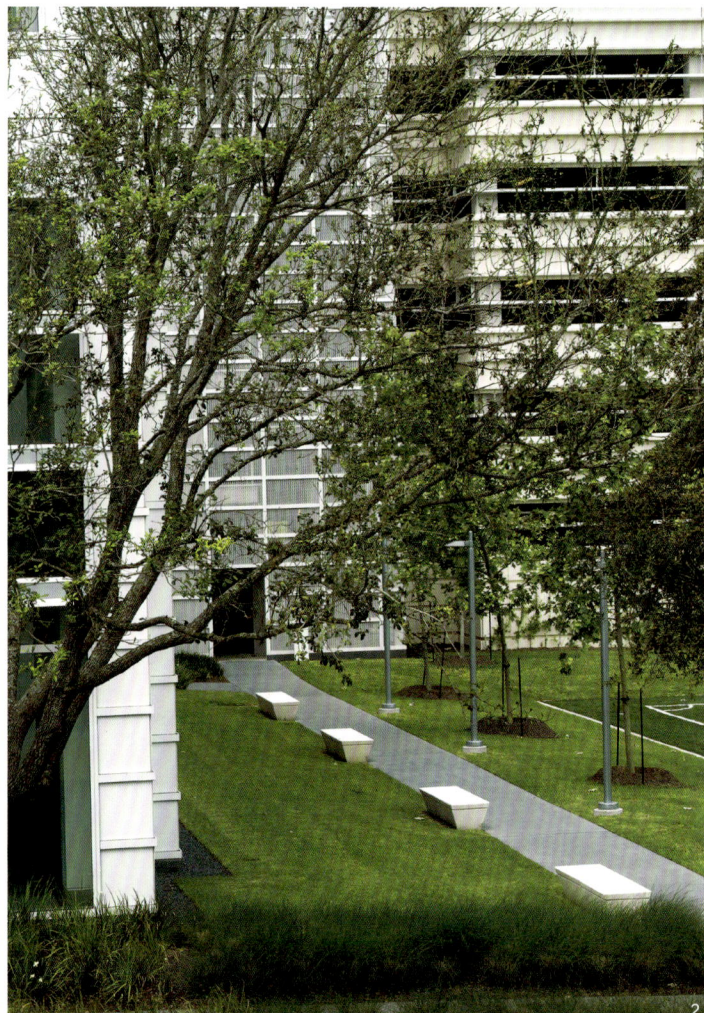

　　该项目位于休斯敦的"能量走廊",占地面积250 905 m²,是世界上最早的企业园区设计典范之一,这种郊区企业园区的模式是凯文·罗奇于1985年提出来的。20年后,康菲石油公司的发展进入了一个关键时期,创新成为其进一步发展和壮大的惟一动力,而园区改造正是体现其创新精神——2020年总规划的第一步。

　　1985年,康菲石油公司从休斯敦市区迁移到休斯敦西郊的Woodcreek园区。该园区在当时的商界和设计领域中被公认为是理想的企业工作环境,不仅便于员工出入和停车,还提供了宜人的自然景观,使康菲石油公司的员工置身于全新的花园式工作环境中。作

家罗伯特·菲什曼在其1987年的著作《资产阶级乌托邦:郊区盛衰》中将这种建筑的格局定义为"科技郊区"。这种设计创意最初是在1979年由罗希·丁克洛公司提出的,设计师在园区内设计了一座三层高的活动场馆,并在周围设置了草坪、绿树和中心湖等。这种设计几乎不用增加户外功能区就已改变了园区的外观,营造出了田园式的景观。

　　20年后,The Office of James Burnett与著名的设计公司Pickard Chilton建筑师事务局、PDR以及Kendall/Heaton公司共同参与了该改造项目。第一阶段是对园区西部的接待中心和健身中心进行规划与施工。设计在园区原有的基础上对接待中心的接待次序进行了调

整,并在园区内新增了庭院、运动场、花园以及沿围墙而建的散步小径。

　　第二阶段的改建计划目标是要设计出能保持20年不落伍的设计,这恰好与康菲石油公司综合发展总规划的时间相吻合,并依照客户提出的要求设计一个足以容纳4000人、配备各种设施的企业园区。该项目的设计着眼于达到"3P标准"即人员(People,吸引和激励有才华的员工)、过程(Process,整合利用最前沿的科技元素)和场所(Place,提升环境的可持续发展)。为了达到这个标准,景观设计师与设计团队的其他成员积极合作最终确立了设计方案,不仅能够改进园区的出入通道、解决机动车与行人在使用通道时的冲突,

而且还将园区周围的庭院作为园区与外界的过渡地带，建造了一系列休息区和户外空间。

这项新的发展规划将设计一系列户外空间，让企业员工可以更好地利用这些绿色空间。为了实现"3P标准"，设计团队还修建了车库、健身中心、诊所、会议室等，并保留了周围原有的橡树，以营造出大片树阴。设计师还将园区内的机动车道按车流特点分成不同类别，花岗岩地面的停车场提供给来访的重要客人，而员工则从专门的员工通道进出；另外，还为外来车辆设置了专门的安检通道。

户外空间主要包括两个原有的位于橡树带的餐厅、一个可以举办公司活动的绿色庭院以及一个设有舞台、灯光和舒适坐椅的花园平台。除了这些户外空间，设计团队还特别在花园里设计了一处私密空间。健身中心被改建成各具特色的三个区域——铺有高质量人造草皮的足球场、一条宁静的花岗石沙砾慢跑道和一个为人们提供思考空间的柏树园。

整个设计方案都采用了防眩光的材料，设计团队还在多处空间中栽种了可以遮阴的树木，并在建筑的入口处及庭院内进行了水景设计。园区内的环路都采用特制材料铺设而成，将各个区域紧密地联系在一起。户外设施主要包括了沿路边设置的固定坐椅和庭院内可移动的坐椅，为人们提供休憩之所。项目中的植栽主要包括了树阴茂密的树木、一些独特的苗圃、在活动区的草坪以及位于园区四周无需修剪的草木。

为了实现并达到人员、过程和场所的"3P标准"，将康菲石油公司总部从一个以机动车为主导的区域转变为以行人为主导的区域，设计团队将人性化的景观设计元素融入到园区内的每一个角落，成功地打造了这个经典的企业园区环境。

The 62–acre World Headquarters of the ConocoPhillips company, located in Houston's "Energy Corridor," is one of the first and finest examples of contemporary corporate campus architecture designed as the model suburban campus by Kevin Roche in 1985. After twenty years, the company reached a critical point where growth and expansion was a necessary step in the re-development of their grounds. This project is the first phase of a campus remodel that implements the goals and objectives outlined in the company's 2020 Master Plan.

The Office of James Burnett, working with the notable firms of Pickard Chilton Architects, PDR and Kendall/Heaton Associates, participated in the planning and execution of this first phase of the re-development process: the Visitor and Fitness Center on the west side of the campus. Improvements included a new visitor arrival sequence, campus courtyard, sports fields, garden courts and a perimeter-walking trail.

In 1985 ConocoPhillips moved from a downtown Houston tower to the newly completed Woodcreek Campus in suburban west Houston. During this time, the suburban corporate campus was viewed within the business and design community as the model environment for the corporate workplace. This campus model allowed for easy employee access, convenient parking and office views to the surrounding landscape. Writing in Bourgeouis Utopias: The Rise and Fall of Suburbia, (1987), Robert Fishman termed this arrangement of buildings the "Technoburb". Here, ConocoPhillips employees found themselves in a new, park-like setting. This original design by Roche Dinkeloo Associates in 1979, placed 3-story pavilions in a landscape comprised of lawn, trees and a central lake. This pastoral landscape offered idyllic and removed views to the exterior grounds, with little outdoor program or functional spaces.

Twenty years later, an expanded design team, comprised

of architects, interior designers and landscape architects, were challenged to update the campus for its next phase, a design which should endure 20 years. The process began with the preparation and completing of the comprehensive 2020 Master Plan. This visionary document included an objective statement from the client that outlined an intention to provide support and amenities appropriate for a 4000-person work environment. The plan focused on meeting three criteria: People (to attract and inspire the best talent); Process (to integrate leading-edge technology); and Place (to promote an enabling environment). To meet each of these goals, the landscape architects, in collaboration with the design team, developed a plan that would improve campus access, separate vehicular pedestrian conflicts, buffer the campus with a park perimeter, and create a series of employee courts and outdoor spaces.

The new development called for a number of programmed exterior spaces to allow the employees to better utilize the abundant exterior green space. The program, based on the three directives, included a new parking garage, fitness center, medical offices, meeting rooms and a new town hall. The buildings were located in an alignment that preserved existing heritage oaks, preserving the generous shade and maturity of the canopy. The new vehicular circulation separated traffic into a distinct arrival sequences, based on program: a granite auto court for important visitors; a series of intelligent entry points for employees; a secure checkpoint for service vehicles.

Green space amenities included: two new dining courts, located under the expanse of the existing Live Oak trees; The Courtyard, a campus green, sized to host campus-wide events; and perimeter garden terraces, developed to accommodate a stage, lighting as well as provide comfortable seating for small groups. In addition to the creation of these large (group) spaces, smaller garden courts were created to provide intimate outdoor rooms. The fitness center program expands outside in three distinct features: a 2/3 size soccer field for intramural competition with a high-performance synthetic turf and European-style goal posts; a serene jogging path winding through the perimeter of the campus, surfaced with decomposed granite gravel; and a lush, bermed garden of cypress trees, intended to serve as a space for contemplation or meditation.

The overall design solution incorporated the use of non-glare materials, shade from tree canopy and arbors, and the incorporation of water at key building entries and courts. Circulation paths were patterned with specialty paving, delineating a hierarchy of connection. Site furnishings included fixed seating along the circulation paths and moveable tables and chairs in the courts, to allow for maximum flexibility. The plant materials used on site include canopy forming shade trees and a balance of special planting beds, lawns in high use areas and un-mowed grasses at the perimeter.

With the aim of transforming the ConocoPhillips World Headquarters from a vehicle-centered entity with a removed, viewed landscape, into a pedestrian-friendly experience, the Office of James Burnett provided design elements that would engage the campus user on a human scale, where people, process, and place combine for a successful campus environment.

1 员工休憩处
2 夏日郁郁葱葱的树木带给人们清凉之感
3 一条小径将办公楼连接在一起
4 运动场

文化的交流与融合 —— 加拿大驻韩国大使馆

Communication and Integration of Two Cultures — Canadian Embassy in Korea

撰文：Tarek El-khatib　　图片提供：Kim Yong Kwan

　　该项目体现了加拿大与韩国两国之间在文化上的碰撞，也突出了二者的共性，尤其表现在尊重自然这一点上。该项目位于首尔市中区贞洞的德寿宫附近，场地中保留了一棵具有520年历史的古树，体现出人们对大自然的尊重。

　　项目所在地的历史背景对建筑材料的选择、场地的布局与设计都产生了重要影响。中区贞洞的多数建筑是由从灰色到红色间不同色度的石材、砖块和木材

建成，这一区域包括德寿宫、学术中心和使馆人员居住区等，具有丰富的历史文化底蕴。因此，大使馆的建筑材料都是经过细心挑选的，如暖色调的砖块和灰色的花岗岩，以便与中区贞洞的其他建筑相协调。设计师在设计时参考了印象派时代的加拿大风景画作品，西侧建筑物仿照了劳伦斯·斯图尔德·哈里斯的作品《罗宾逊山》的风格；而东侧建筑物外立面的设计则综合了韩国传统的布艺和加拿大印象派大师汤姆·汤姆

森的作品《Evening Canoe Lake》的风格，营造出自然和谐的美感。

设计突出了入口处的古树的重要性。如今，这棵古树已成为大使馆入口庭院的标志。在建筑和古树间留有一定距离，这样既拓宽了环绕古树的人行道，又能使行人更多地体验到在树下乘凉的乐趣。建筑物外围的木质隔板营造出静谧的氛围，这些隔板由加拿大红雪松木制成，参考了韩国木质隔板的样式，并与德寿宫的围墙相协调，此处设计使大使馆周围空间的可视性与私密性达到了平衡。对古树的保护是设计成功的关键，设计师尽可能地降低施工对古树根部的影响——在施工期间将古树周围的区域进行围护，保证古树生长所需的地下水。景观设计师 A.D.Regehr 提议由植物学家来制定保护古树的方案，因此，邀请了 Sang Yong Nam 教授和 Kyeong Jun Lee 教授负责古树的保护工作。

一堵混凝土墙矗立于使馆外的庭院中，将人们的视线引向喷泉和花园。由于大使馆已经开放，因此这里也成为了这一地区的焦点和拍照的最佳景点。这处

1　休息和观赏古树的安静空间
2　混凝土墙和入口庭院处的长椅

温暖且受人欢迎的庭院象征着加拿大和韩国的两国关系和睦融洽。

在半开放的空间（如接待中心、展厅和餐厅等）以及设有可伸缩屏风的多功能厅中，人们可以透过木质隔板隐隐约约地欣赏到花园。花园被围墙和树阴围合起来，并设置了水景。在这个布局紧凑的场地上，线性花园是设计的一个亮点。花岗岩铺装地面由入口庭院开始，经过主建筑区一直延伸到花园里，好似"浮动的原木"。

设计师采用多种设计手法，尽可能地实现场地的可持续性，如充分利用已有场地、直接将排水系统引至景观区、搭配使用当地材料、通过高密度的自动化系统减少停车场空间、提供多处开放空间、设计隔板和幕墙以减少光污染等。设计师在克服了一系列的技术难题后，最终营造出尊重大自然并拥有如"绿洲"般舒适的空间。该项目的设计在表达对韩国人民和韩国文化的赞赏之情的同时，也成功地体现出加拿大文化的特点。

The design of the Canadian Embassy in Seoul creates a dialogue between Korean and Canadian cultures expressing common links, in particular, a shared reverence for nature. This unique site shares a "place" with a 520-year-old tree, a living symbol of nature, called Hakjasu or "scholar" tree in the historic Jeong-dong district near Deoksoo Palace.

The building is located in the historic cultural district that combines the Deoksoo Palace, academic centres, and ambassador residences. This historic Jeong-dong context plays an important part in the building materials, organization of the embassy and site design. Jeong-dong is built of stone, brick and wood in varying hues ranging from grey to red. Embassy materials such as warm toned brick and grey granite were chosen to carefully harmonize with these colours and textures and to extend the pedestrian walk that meanders along the undulating Deoksoo Palace wall. Canadian references inspired by impressionist images of the Canadian landscape were introduced in the massing and skin of the building. The west block is a majestic and simple form in the tradition of Lawren S. Harris' "Mount Robinson", while the east block creates a natural rhythm against the sky inspired by both patterns found in traditional Korean textiles and the Canadian impressionist, Tom Thomson's "Evening Canoe Lake".

We chose to heighten the importance of the ancient 500-year old tree by placing the building entrance at this location. This Hakjasu tree now forms the focal point of the embassy's entrance plaza. The building was pulled back from the tree so that the initial sidewalk, which ran tightly beside the tree, was widened and curved to provide a contemplative setting for the tree allowing pedestrians to experience the full majesty of the tree canopy. Horizontal wood panels wrap the base of the building creating a calmness and serenity.

These slats made of Canadian western-red cedar reference Korean wooden screens as well as the undulating walls of the Deoksoo Palace. As one enters the embassy the screen provides just the right balance of visibility and privacy. Developing a protection plan for the tree was critical to the success of this project. The massing was designed for minimum impact on the tree root-ball and for maintenance of groundwater to the tree during construction by tanking a large area around the tree. Landscape Architect, A. D. Regehr proposed the initial tree maintenance plan for consideration by the tree specialists. During construction Professors Sang Yong Nam and Kyeong Jun Lee monitored the health of the tree.

A concrete wall anchors the plaza and draws views to the picturesque fountain and the walled garden beyond. Since the building has opened, the plaza has become a neighbourhood focal point and a photo op destination. This warm and welcoming entrance courtyard expresses the spirit of the relationship between Korea and Canada. The design successfully represents Canada while expressing an appreciation of the people of Korea and their culture.

At ground level the semi-public spaces (reception, exhibition, dining) offer subtle views through wood screening to the exterior garden. The flexible multi-purpose room with retractable walls allows visual expansion into the private walled garden with water elements and tree canopy. The colour and texture of the garden wall will develop as the vines start climbing the vertical cables located at intervals.

On this tight site the linear walled garden with reflecting pool was a key design feature. The granite paving forms a carpet resembling "floating logs" in the horizontal plane that begins at the entrance courtyard, continues through the main spaces and ends in the garden.

Using an existing urban site, directing drainage into landscaped areas, coordinating with local materials, minimizing parking through a high density automated system, providing open space, reducing light pollution by screening and curtain wall design—these were all measures taken to promote sustainability. In the end technical challenges disappear and the impression of the site is a sense of ease and welcome around the central tree and a jewel-like oasis within the walled garden.

绿色办公空间设计的典范 —— 丹麦无线电台景观设计

A Classic of Green Business Space — Landscape Design of Danish Radio

撰文 / 图片提供：Pedro F Marcelino C. F. Møller Architects　　翻译：武秀伟

丹麦无线电台是丹麦最大、最古老的电子传媒企业之一，它创立于1925年，是一所公共服务机构，位于奥胡斯的广播大厦园区内。

奥胡斯广播大厦园区建于1975年～1981年，园区内建筑低矮密集。根据建筑物的大小、形状就能够反映出其功能，根据其各自的功能，园区内设置了不同的电缆，如无线电电缆、办公电缆、电视电缆和工业用电缆。这些电缆由贯穿东西南北的两条大道连接在一起，穿过广场，通向两个大型工作室。

东西走向的大道直接与主入口相连。园区的附属单位，如档案馆和印刷室等也与这条大道连接在一起。几条小路从大道两侧延伸出来，路边设有一些专门的休息区，与天井和内部庭院连在一起。

广播室位于园区中央的综合楼的上方，人们很容易从外部看到广播室的轮廓，因为园区内的绝大部分建筑只有1层～2层楼高。整个园区坐落在一处高地上，规则的小石板铺砌的堤岸将其围合起来，以缓解地势的高差变化。

Danmarks Radio is Denmark's oldest and largest electronic media enterprise. It was founded in 1925 as a public service organisation. Parts of Danmark Radio's activities are situated in a broadcasting campus in Aarhus.

The campus was built in 1975-81 as a low, dense, urban structure where the individual functions are reflected in the size and shape of the individual buildings. Functions that naturally belong together are situated in separate "lines": a radio line, an office line, a TV line and an industrial line. These functions are interconnected by two main thoroughfares north/south and east/west. Where these lines cross a foyer/square is naturally formed, giving access to the two large studios.

The east/west thoroughfare is in direct contact with the main entrance. In connection to the thoroughfare the outward functions for the whole campus are situated: record library, printing office etc. From these thoroughfares smaller streets radiates in a dense pattern. Along the streets rest areas are placed deliberately and they are in contact with patios and inner courtyards as an extension the streets and rest areas.

From the outside it is easy to "read" the Broadcasting House's structure in its profile: Most of the complex is one or two stories high. Above this rises the studios centrally situated in the campus and at the main entrance. Furthermore, the campus is situated on a plateau bounded by a precise and softly shaped bank of small sett paving, which absorbs the movements of the terrain.

典型的澳洲风情 —— 悉尼瑞斯迈公司园区

A Distinctly Australian Flavour — ResMed Corporate Campus, Sydney

撰文：易道公司　　改编：Pedro F Marcelino　　图片提供：易道公司 David Lloyd Simon Kenny

在易道公司的巧妙规划下，一片退化凌乱的溪边浅滩摇身一变成为了瑞斯迈（ResMed）公司在澳大利亚悉尼的一个园区。该园区规划统一，占地面积为120 000 ㎡。以项目原址的生态环境和历史作为框架背景，设计的理念是将这条溪流看作"思想的溪流"，将该项目的公众性、娱乐性、美学、环境和雨水处理统一起来。

该项目的客户是澳大利亚的上市公司瑞斯迈，专门开发、生产和销售用于诊查、治疗和长期应对睡眠呼吸障碍与其他呼吸障碍的产品，目前业务遍及全球68个国家。瑞斯迈要求设计方为公司新建的位于悉尼诺威斯特工业园全球总部提供一体化的景观设计与环境管理服务，包括景观设计、项目总体规划和项目全程施工监理等。

虽然瑞斯迈的主要市场在美国和欧洲，但是所有的生产都是在位于诺威斯特工业园的总部完成的。为了突出该工业园对公司的重要性，客户坚决要求设计带有浓郁的澳洲风情；此外，景观设计要突出"工作场所"的特色，既要满足公司总部1200名员工在此开会的需求，还可以满足在休息时间开展娱乐活动的需求；设计还必须有足够的创意，要充分尊重澳大利亚的文化历史和地区生态环境。

该项目的总体规划在视觉上分为多个层面，但是
又有着千丝万缕的联系。项目所在地靠近澳大利亚最
著名的文化遗产名胜地——伊丽莎白·麦克阿瑟庄园，
如此深厚的历史背景给该项目烙上了深深的文化印
记。这一古老庄园是将绵羊引进澳洲的先行者约翰·
麦克阿瑟和伊丽莎白·麦克阿瑟的故居，古色古香的
房屋的大部分现在仍保存完整，是澳大利亚殖民时期
留存下来的最古老的欧式建筑。庄园位于澳洲原住民
Dharug 人的 Burramattagal 部落的居住区，从而使庄园
更具有象征性。早在 19 世纪初期，庄园曾被修缮加固，
周围有花园、果园并且开辟出一片林地。

与这样一处充满文化气息的历史名胜为邻，该项
目的设计基调或多或少也会受到一些影响。设计既满
足了可供 1200 人学习的功能要求，同时又提供了适合
濒临消失的页岩平原林地生长的生态环境（目前在悉
尼城区周边仅存 13% 的原生林地）。总体规划依托于
一条中央隆起带，使园区呈东西走向。被设计师亲切
地称为"拉链"的隆起带以水的形式突出其自身的存在，
连接几处永久性水体的溪流在穿越场地的过程中重塑
了形态。这一中央水元素代表的就是上文所提及的"思
想的溪流"，或者是易道公司定义的意识流——永无止
境、动态十足、随时代而变迁。

小溪中的水源于蓄存的雨水，流经场地内一系列
不规则排列的池塘和堤坝，就像横跨在小溪上的树，
造就了阶梯式的变化。设计师在每一个水流的交汇处
都做了精心设计，比如具有独特创意的艺术小品或装
置，它们在瞬间改变了人们对场地的认识和第一印象。
小溪形态的变化和在场地内的布局正是为了让人们缅
怀这片土地的历史：从伊丽莎白·麦克阿瑟庄园附近
的锈铁和原木，到当年囚徒们构筑的砖墙，直至现代
高科技工业园区内闪亮的金属与玻璃建筑。

What was once a degraded and intermittent creek has become the central unifying element of ResMed's new 12-hectare campus in Sydney, Australia. Wrapped in a framework of site ecologies and histories, EDAW's concept envisages the creek as an "Ideas Stream", a flow of waking thought, unifying the site's corporate, recreational, aesthetic, environmental and stormwater treatment objectives.

EDAW projected the site for Australian-owned and publicly-listed ResMed corporation, a reputed player in the Sydney and New York City stock markets. As a specialized developer, manufacturer and marketer of products for the screening, treatment and long-term management of sleep-disordered breathing and other respiratory disorders, ResMed has attained worldwide projection, currently operating in 68 countries around the world. EDAW were engaged by ResMed to provide a full landscape architectural and environmental management service for their proposed new 12-hectare international headquarters in Sydney's Norwest Business Park. The EDAW team was responsible for work ranging from landscape and site master planning through to construction supervision for all stages of the project, in a highly significant campus development with international repercussions.

Although ResMed's main markets are in the United States and Europe, all the manufacturing is completed at the corporate headquarters within the Norwest Business Park. In order to showcase the importance of this site for the company, the client was adamant on a distinctly Australian response to the brief. Additionally, throughout the briefing process, ResMed required EDAW to provide a site response that encouraged the landscape's usage as a workplace, where many of the 1200 staff members could conduct meetings and recreate during down time. The landscape was also required to be innovative and respectful of both Australia's cultural history and the ecology that supports the area.

The landscape master planning response was multi-layered in its vision and cohesive in its execution. The plan recognises the cultural importance of being located next to one of Australia's most significant heritage sites, Elizabeth MacArthur Farm. The historic estate in Parramatta, in Sydney's suburbs, was the home of the wool pioneers John and Elizabeth MacArthur. The old cottage, which remains mostly intact, is currently the oldest surviving European dwelling from Australia's colonial period. It is set in the aboriginal lands of the Burramattagal clan of the Dharug people, lending it further symbolism. In the early 19th century, the original cottage was upgraded and surrounded by gardens, orchards and acres of cleared bushland.

This respectful neighbour partially framed EDAW's approach to the project. It meets the functional requirements of a 1,200-person learning campus, while providing catchment-wide ecological benefits befitting a site with a remnant stand of the endangered Shale Plains Woodland (currently, only 13% of the original woodlands surrounding Sydney's urban sprawl survive). The entire site master plan hangs off a central spine, which bisects it in an east-west direction. The central spine in the landscape—or "zipper", as the designers dubbed it—manifests itself in the form of water. A stream connects several permanent water bodies and

reinvents itself as it moves down through the site. This central water element represents the "Ideas Stream", or stream of consciousness conceptualized by EDAW—infinite in nature, and dynamic enough to change with the company over time.

The stream is fed by harvested stormwater, descending logically through a sequence of ponds and dams that fall randomly across the site, as trees fall across a forest stream, producing a series of stepped level changes. Each crossing intervenes, leading paradoxically away from the stream, taking form as some folly (or idea). These follies are art installations or pieces of landscape architecture by their own merits, performing the function of momentarily shifting both the participant's perception of the site and his/her immediate thinking. Changes in the stream's materiality and textures are woven throughout the area to recall site memories and histories: of the rusty metal and timber of nearby Elizabeth MacArthur Farm, of the bricks of the once local convict-staffed brick works, to the polished steel and glass architecture of today's high-tech industry in the business park.

新加坡樟宜国际机场T3航站楼

Singapore's Changi Terminal 3

撰文：Franklin Po Sui Seng　图片提供：Albert K.S.Lim　翻译：董桂宏

新加坡樟宜国际机场的T3航站楼内部悬挂着一幅巨大的绿色"织锦"——它由垂直生长的植物组成，大气磅礴，让人们自然联想到东南亚的热带雨林。巨幅"织锦"不仅柔化了这座巨穴式的工业建筑，还将整座航站楼划分为陆侧和空侧两个区域，并且与登机区域的垂直空间相连接（T3航站楼与登机区域通过玻璃安全屏分隔开来）。

原有的花岗岩分隔墙被改造成一幅巨大的绿色"织锦"，攀缘植物和气生植物郁郁葱葱，形成了壮观的垂直花园景观，成为了T3航站楼最主要的特色。"织锦"高15m，覆盖了整个场地的四分之三（航站楼全长400m，其中"织锦"长300m），并起到了按比例缩小巨大内部空间的作用。

绿色"织锦"展现了富饶多样的东南亚热带雨林风情——创意新颖、个性十足、充满震撼力。它可以通过支架背部的T形通道进行维护。植物被预先种植在靠近不锈钢缆绳的植栽容器中，不需要任何扣件便可将缆绳固定在框架上。

设计施工

四年多来，景观设计师、园艺师和建筑师定期对该项目进行探讨，共同研究如何设计硬质景观和软质景观，并对一些设计细节和施工情况等进行深入探讨。设计团队从马来西亚和泰国进口了多种植物，并在种植之

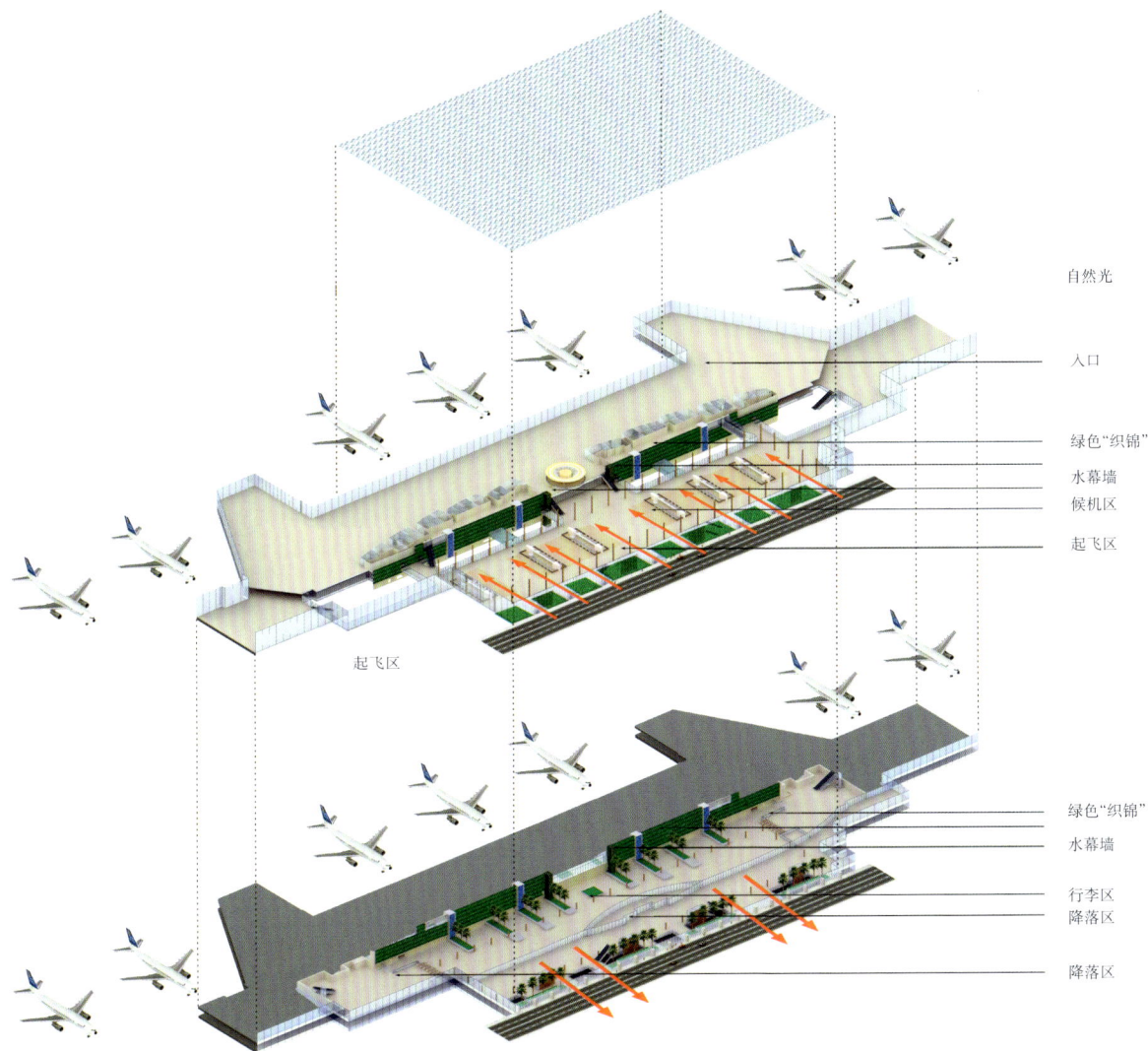

自然光

入口

绿色"织锦"

水幕墙

候机区

起飞区

起飞区

绿色"织锦"

水幕墙

行李区

降落区

降落区

前对其进行了特殊护理，如病虫害预防等，以确保植物能够在室内茁壮生长。在施工过程中，绿色"织锦"首先完工，但在随后进行其他施工的时候，长期空气流通不畅和灰尘污染曾一度导致绿色"织锦"上的植物毫无生气（这也是设计上的一处失误）。

屋顶天窗的设计确保了绿色"织锦"上的每一棵植物都能够最大限度地进行光合作用；水滴灌溉系统保证了植物生长所需的水分；而手动喷雾装置则具有减少灰尘堆积和增加枝叶营养的双重功能。整套系统都在 T1 航站楼经过长达四五年的测试，设计团队利用这段时间充分研究了室内光照水平，并且逐渐选择出适宜的植物。

影响

T3 航站楼的室内环境舒适宜人（已通过多项严格指标测试），巨幅绿色"织锦"有效地改善了室内的空气质量，使旅客心旷神怡。由于航站楼内来自世界各地的旅客川流不息，可能导致室内植物受到病虫害的污染。因此，设计师要确保航站楼内不含任何有毒化学物质，于是采用非毒性病虫害的防治方法来长期维护航站楼的景观。

巨幅绿色"织锦"不仅具有装饰作用，更彰显出景观的自然性。在 T3 航站楼这样巨穴式的建筑中，景观要同时具有环保作用，人性化的设计改善了室内环境，使景观与环境和谐统一。

可持续性

T3 航站楼属于人造景观，设计需要兼顾多个环节，如屋顶天窗既要保证能够吸收更多的自然光，又要保证充足的光照（紫外线和红外线）能够进行光合作用；喷雾系统用来维持室内湿度；养分则由滴灌施肥进行补充；预先培植的植物用以替换枯萎的植物；设计师选用大型的植栽容器，容器内土壤充足，能够满足长期种植高大橡树和茂密灌木丛的需要。

Large-scale vertical planting evoking a South East Asian equatorial rainforest was introduced into the interior of Singapore's Changi Terminal 3 to structure and soften an otherwise cavernous industrial building. A woven tapestry of living plants not only divides the mega-building in plan into landside/airside sections but also connect the vertical space of the check-in/arrival areas, which are separated by a glass security screen.

The granite-clad dividing wall was transformed into a large green wall, a vertical tapestry of climbing and aerial plants. This vertical garden is the building's most distinctive feature, spanning 1000 feet of the 1300 feet length building and at 50 feet high, helped to scale down the high interior space. The green tapestry exhibits the richness, diversity and character of the Southeast Asian Equatorial Rain Forest. Its bold, dominant form creates a powerful identity for the terminal's vast space. The green tapestry is designed to be maintained easily from catwalks behind the suspended structure. The plants are pre-grown in containers on stainless steel cables which are secured to the structural frame without need of fasteners.

Design coordination and implementation

The design team of landscape architects/horticulturists/architect, over a period of four years, attended to the project on a regular basis, reviewing hardscape and softscape design, details and performance of the system. Portions of the tapestry were completed while construction was still in progress.

The results for plant establishment were initially poor due to constant dust and poor air movement. Plants were imported from neighboring Malaysia and Thailand and allowed to be acclimated to similar conditions before installation. Pest control was monitored by ensuring that all plant specimens brought into the interior was free from infestation.

Since natural light is available, the skylights were adjusted to allow maximum light to reach all portions of the green panels. A drip irrigation system is designed to water and fertilize the plant specimens; hand misting is also carried out performing a twofold function: minimizing collection of dust on leaves and foliage feeding. Mock-ups were tested over a period of 4 to 5 years at Terminal 1 which allowed the team to study lighting levels and selection of plant species for optimal performance.

Impact

Terminal 3 is an interior environment subject to stringent controls for human comfort. The plants contribute a visually cooling and refreshing ambience, but do not in any significant way alter the environment of the interior. Pest infestation could become a problem as there is a continuous influx of people and ornamental plants into the terminal. Maintenance is a continuous and long-term issue. The client being sensitive to pest control through non-toxic methods is ensuring that the interior environment is free from noxious and toxic chemicals.

The green tapestry is not merely a decorative accent but a bold environmental statement complimenting the architecture, demonstrating that landscape as architecture is capable of humanizing the interior environment of an enormous and complex building.

Sustainability

As part of an artificial, man-made environment, the Terminal 3 landscape requires continual monitoring. Light (U.V. and infra-red) levels must be sufficient for photosynthesis, and is available from the roof skylights. Humidity is replenished by a misting system and nutrients by drip fertigation. Weak plants are removed and replaced with pre-grown substitutes. The baggage island planters and planting platforms are not individual tree planters, but huge planting troughs with sufficient soil to support long-term life-span of tall palms and shrub groupings.

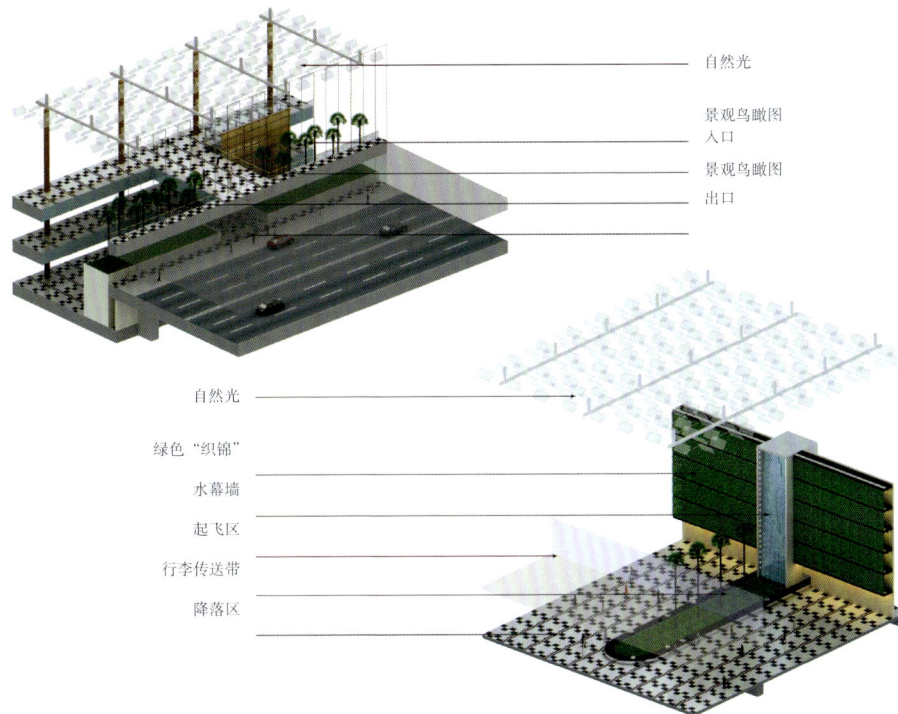

自然光

景观鸟瞰图
入口

景观鸟瞰图
出口

自然光

绿色"织锦"

水幕墙

起飞区

行李传送带

降落区

慕尼黑Highlight商务大厦

Highlight Towers, Munich

撰文 / 图片提供：Rainer Schmidt Landscape Architects　　翻译：申为军

该项目位于慕尼黑市施瓦宾区北部，两座大厦分别高 113m 和 126m（28 层和 33 层）。该大厦是慕尼黑市最高的建筑物之一，是城市重要的标志物。

于 2004 年竣工的这两座双子大厦现如今已经成为了慕尼黑现代建筑的典范。为纪念两位建筑界的大师，双子大厦周边的新街道被特意命名为路德维希·密斯·凡德罗大街（Ludwig Mies van der Rohe Strasse）和瓦尔特·格罗皮乌斯大街（Walter Gropius Strasse）。该项目的设计师赫尔穆特·杨（Helmut Jahn）曾在芝加哥师从于密斯·凡德罗，并深受其影响。

从奥登广场 (Odeonsplatz) 出发，沿着路德维希大街这条历史悠久的城市中轴线直行，穿过凯旋门（Sieges

1 在大厦最高层的俯视图
2 双子大厦仰视图
3 道路景观
4 自行车停放处

Tor）稍向北走就可以抵达双子大厦。对于沿 A9高速公路南行的游客来说，双子大厦是一个显著的城市界标。

双子大厦由两座玻璃桥相连，站在桥上可以欣赏到慕尼黑全景和阿尔卑斯山雄伟壮观的美景。大厦旁边的两栋楼房是一家酒店，负责为大厦中其他公司提供食宿服务。写字楼总面积为 70 000m²。

大厦平滑的玻璃外墙给人以明亮、通透的感觉，每一间办公室都采光充足、温度适宜。夜晚，华灯初上，

大厦各层的蓝色地灯一起点亮，清晰地勾勒出双子大厦的轮廓。

除了玻璃幕墙结构，双子大厦的另一独特景观特色是其清晰划分出的若干区域。这些景观区虽然大小不同、形状各异，但都与大厦整体非常协调。深灰色花岗岩与白色花岗岩石条交替构成大大的同心圆，在两座大厦之间形成中心广场，穿过大厦的前厅一直延伸至大厦的另一侧。从楼上俯瞰，这些同心圆就像一块漂亮的地

毯，又或者是一个承托着两座高塔的精致托盘。另一个稍小一些的同心圆图案则置于酒店门前，作为停车带。

修剪整齐的树墙和长长的坐椅等直线元素让大厦之间的中心广场充满了生机和活力。巨大的假山看起来好似位于高楼大厦之间的下沉区，既加强了对比效果，又与高大的楼房相得益彰。大厦周围栽种的植物专门选取了轮廓分明的品种。大型种植区选种单一品种，有效地强化了景观的图形效果。

The Highlight Towers in the north of Schwabing with the twin towers of 113 and 126 m (28 and 33 floors) being one of the highest buildings in Munich they add a significant accent to the Munich skyline.

Completed in 2004, the towers are an example of new architecture in Munich. The streets forming the new address are well chosen and not by chance called Ludwig Mies van der Rohe and Walter Gropius Strasse, both were pioneers in the architecture and building industry. The architect Helmut Jahn studied under Mies van der Rohe and was greatly influenced by him in Chicago.

The towers are slightly offset from the main historic axis from Odeonsplatz along the Ludwig Strasse, through the Sieges Tor towards the north. The project is also a landmark for travelers and visitors coming along the A9 Autobahn from the north.

Two glass bridges connecting the two towers give tenants breath-taking views over the city and the Alps. Two lower buildings surrounding the towers accommodate a hotel, catering for other office facilities. In total there 70,000

square meters office space.

The flat and smooth façade appear light and transparent offering an optimal office lighting and climate. At night, blue lights marking the corners of the building on each floor accentuate the form.

In contrast to the glass architecture the landscape features are divided into various zones that in size and graphic are in scale with the building. Large concentric circles generated from between the towers form a central plaza and continue through the tower lobbies and exit the opposite façade. Alternating dark gray and white granite bands strengthen the graphic. From above this forms a type of carpet or serving tablet on which the towers are placed. A second, smaller circle forms the drop-off for the hotel.

Linear elements such as cut hedges and seating benches strengthen the dynamic into, through and out of the plaza between the two towers. Oversize mounding looking like sunken spheres adds to the contrast and is yet in scale with the large buildings. The planting selection is simple and clear. Large planting areas using the same species with each element is effective in strengthening the graphic effect.

加州捐赠基金会景观设计

The Landscape Design of the California Endowment

撰文：Samantha Harris　　　图片提供：Tom Bonner　Lisa Romerein　　　翻译：武秀伟

该项目的景观设计与加州捐赠基金会总部的设计风格相似，体现了基金会所肩负的责任与使命。该设计再现了加州地理面貌和植物种类的多样性，恰如其分地体现了人们健康的生活方式，并与周围的文化机构建立起友好的互动关系。该项目整体结构包括一座1490m²的庭院、一个大型的停车场和建筑物外围的开阔空间。

加州各地的代表性植物和石雕作品都会聚于此，可见该基金会在该州受到的拥护程度。植物配置包含4类树种、8种草本植物、30种多年生植物，分别种在26300m²的场地中，以体现加州生态系统的多样性。场地四周分布着5座海岸红杉花园，地面上种着蕨类植物和叶蓟属植物。每座花园中分别种植着6棵大小不同的北美红杉，象征着自然生长的过程。它们雄伟高大，成为具有传统风格的自然景观，与城市的工业背景形成了鲜明的对比。

呈东西走向的胡椒树群沿场地的南北边缘种植着，雕塑般低垂的树木使人们联想起这座大厦赋予这些花园的使命感，在树与树之间还种植着一些低矮的耐旱植物。

Alameda街道前方种植着成排的悬铃木，地面上则种植本地的灯心草和耐旱草种，将这里装扮成宜人的绿色通道，供人们通行，同时该通道将洛杉矶市中心及康菲尔德国家公园连在一起。海滨树种与其周围的植物使人联想起附近的洛杉矶河以及废弃的桑哈·马德烈排水管（曾经是洛杉矶旧城的排水系统）。设计师在新的洛杉矶排水系统周围种植着大量的植物，将植物群作为Alameda街道的一部分景观。悬铃木及其周围的草地一直延伸到停车场，并成为那里的主体景观。

该项目的整体与细节设计都充分展现了加州的人文历史景观。花园墙体的设计贯穿整个场地，从加州各地精选出的石块各具特色，如派因溪的金溪岩、海岸山脉的绿片岩和中央峡谷的萨克拉门托河岩。

Alameda街道对面的花园式庭院面积为1490m²，庭院四周是该基金会的会议室，将庭院与周围的城市环境相融合，为不同的活动与聚会提供了多功能的空间。会议室安装了可伸缩的玻璃墙，使人们可以自由进出，充分享受加州户外柔和的暖风。

受加州旧教区庭院和附近日式花园的影响，该庭院属于当代田园风情式风格。除此之外，加州传统的户外生活方式也对该设计产生了影响。庭院的路面由金色花岗岩和绿片岩铺设而成，路面一直延伸到内部，将室内外空间连接在一起，彩色的条状混凝土路面给人以轻快的感觉，甚至会忽略它下面是一座地下车库。

庭院里有三个矩形的混凝土花坛，里面种植着柳树，并设有内嵌悬臂式木质坐椅，为人们提供了阴凉的休息处。中间的花坛还兼作一处水景，底部的鹅卵石是从萨克拉门托河运来的。庭院周围的栅栏采用当

1　会议中心周围的院落
2　根据附近火车颜色设计的彩色金属条纹

红杉
加州胡椒树
加州悬铃木
红杉
加州胡椒树
红杉
加州悬铃木
行政楼
绿片岩花园墙体
金色花岗岩路面
室内庭院
中央庭院
主路
应急车辆回车场
公共空间
绿片岩路面
绿片岩花园墙体
红杉
药草园
红杉
公共空间
加州胡椒树
Alameda 街
Bauchet 街
N
景观规划图

地随处可见的三角花的藤蔓制成，它们被修剪得几乎看不出原貌，与会议室明亮的色彩交融在一起，显得十分自然。竹篱把庭院分隔成几块小花园，很多加州的花园都采用这种竹篱分隔法进行设计。

一座药草园掩映在南边的角落里，从场地的多个方向都可以进入到该花园。园内种植了八种药草，包括天然芦荟、薄荷和鼠尾草。设计师在与加州捐赠基金会讨论后决定把这座药草园作为一处灵活多变的景观展示给人们，在里面种植不同种类的药草，如从邻近的唐人街移植一些具有药用价值的中草药。设计师希望通过药草园来强化加州捐赠基金会对健康教育的责任感，同时也为员工和参观者提供一处可以放松沉思的场所。

1　建筑的形状与颜色与附近的景色很协调

2　入口处花园

3　入口处的景观

4　植被葱郁的停车场

5　加州的本地植物品种构成了加州特有的景观

6　通向庭院的大型玻璃门

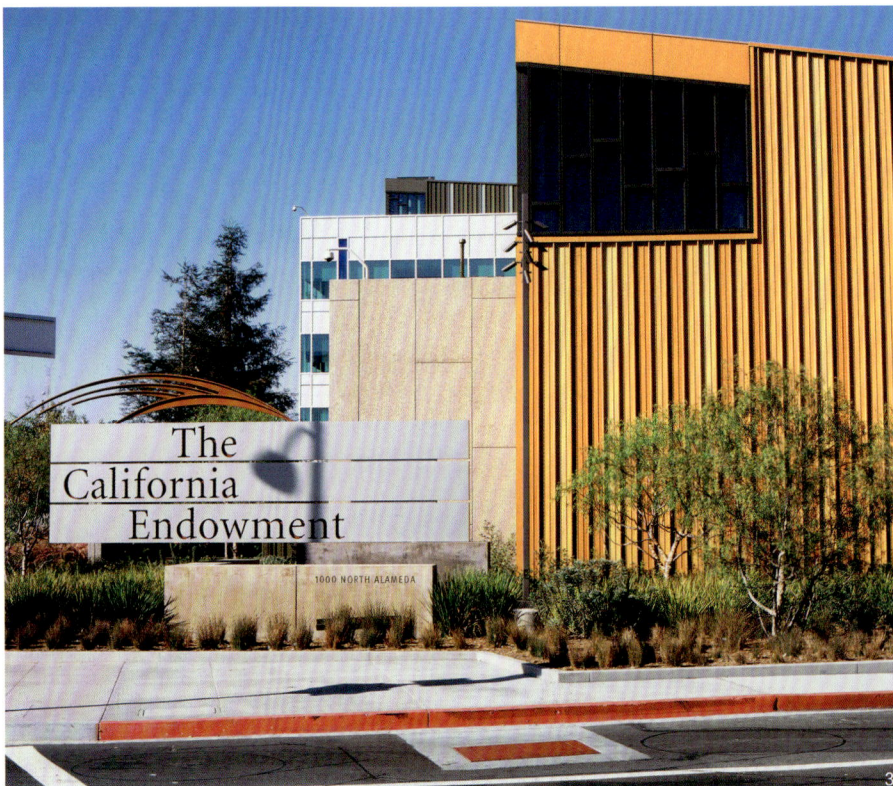

The California Endowment

1000 NORTH ALAMEDA

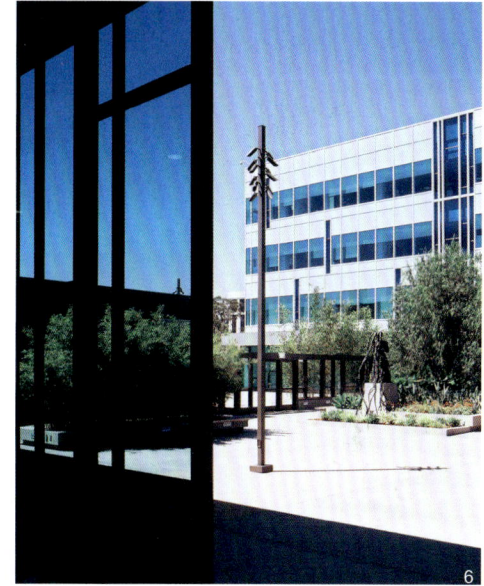

Similar to the architectural design of the California Endowment headquarters, the landscape architecture was designed to embody the mission and principles of the California Endowment. The landscape achieves this by evoking the diversity of the state's flora and geography; capturing the healthy, indoor-outdoor California lifestyle; and creating strong connections with the vibrant cultural enclaves that surround the site. The exterior grounds include a 16,000-square-foot courtyard, a large surface parking lot, and open spaces along the building perimeter.

To symbolize the California Endowment's statewide constituency, the landscape weaves together a diverse range of flora and stonework that are iconic to regions across California. Included in the design are four tree species, eight types of herbs, and 30 types of perennials, which are planted around the 6.5-acre campus in a manner that evokes the state's many ecosystems. Scattered throughout the site are five Coast Redwood Gardens, which are each planted with six Sequoia sempervirens (Aptos Blue) and under planted with fern species and Acanthus. The redwoods are planted at different sizes to emulate natural forest growth. Due to their large size, the trees become iconic natural objects juxtaposed against the site's urban industrial backdrop.

California Pepper Trees are organized along the site's

north and south borders in an east-west direction. The weeping sculptural form of the trees recalls the gardens of the state's historic Missions. Between the trees are drought-tolerant under plantings laid out in a contemporary banded pattern.

Along the front of the campus at Alameda Street, rows of California Sycamores are under planted with native rushes and drought-tolerant grasses, creating a pedestrian-friendly green corridor that makes connections to downtown Los Angeles' civic centers, in addition to Cornfield State Park. The riparian trees and under plantings allude to the nearby LA River, as well as to the now defunct Zanja Madre—the historic irrigation ditch that once served Pueblo de Los Angeles. Rios Clementi Hale Studios imagined the rich flora that may have accompanied Los Angeles' current channelized water systems, and reinterpreted the flora as part of the Alameda Street corridor streetscape. The California Sycamores and grass plantings were extended into the campus' surface parking lot where they are the predominant landscape type.

The people, history, and landscape of California are richly represented through the site design and detailing. Garden walls are located throughout the site and feature distinctive stones culled from regions across California, including Golden Creek Granite from Pine Creek, Green Schist from the Coastal Range, and Sacramento River Rock from the Central Valley.

The 16,000-square-foot garden courtyard faces Alameda Street and is surrounded by The California Endowment's meeting rooms. The courtyard plays an important role in establishing connections to the surrounding urban environment, as well as providing multi-functional spaces that can be programmed for diverse functions and gatherings. In the spirit of California's breezy outdoor lifestyle, the meeting rooms have retractable glass walls that can be moved to allow people to spill outside, flowing between the indoors and outdoors.

The idyllic, contemporary courtyard was influenced by the state's historic Mission courtyards, Japanese gardens from nearby Little Tokyo, and the iconic California outdoor lifestyle. Designed to belie its location above a subterranean parking garage, the courtyard is light and airy with bands of colored concrete paving that are seeded with Golden Granite and Green Schist aggregate. The bands run across the courtyard and extend into the interior spaces, creating a continuity between the indoors and outdoors.

The courtyard features three, rectangular concrete planters containing Geijera parviflora trees and built-in, cantilevered ipe-wood seating, providing a cool shaded seating area at the heart of the garden. The center planter doubles as a serene water feature, and has Sacramento River Rock cobble at the bottom. The Bougainvillea vine—ubiquitous in California residential planting—is applied to security fences, rendering them nearly invisible and providing natural color at a pedestrian scale. The Bougainvillea further relates to the bright colors of the one-story meeting buildings. Bamboo hedges found in many California backyard gardens are used to break the courtyard into garden rooms of smaller scale.

A medical garden is tucked in the southern corner of the site, and is accessible from various points on the campus. It is planted with eight herb types, including Aloe Vera, Pennyroyal, and Sages. The California Endowment uses this as a flexible demonstration garden that could be planted with different medicinal herbs in the future, such as Chinese Medicinal Herbs from neighboring Chinatown. A medicinal garden reinforces the California Endowment's commitment to health education, while providing a stimulating, meditative area for employees and visitors to enjoy.

静谧的都市庭院景观

The Serenity of Pinnacle Urban Courtyard

撰文 / 图片提供：Koch Landscape Architecture　　翻译：张璐

　　该项目位于俄勒冈州波特兰市"河流区"的北部，周围建有两栋高层商务办公楼。项目东边是维拉姆特河，西北则是费尔德斯公园。庭院的设计独具匠心，既为过往行人提供了便利，又在一定程度上营造出一处私密的空间——静谧的都市庭院景观。

　　该项目包括一个自然的生态屋顶、弧形墙和人行道。设计师对弧形墙、路灯以及植栽进行了整体的规划，使其能与庭院中的人行道相得益彰。呈几何形种植的树木将庭院遮挡起来，其颜色和布局也与周围的建筑相互呼应，营造出一处私密的空间，置身其中，人们可以尽享这一空间的舒适和惬意。玫瑰红的紫荆花花瓣点缀着灰色的墙面，桦树与凉亭遥相呼应，紫色的蛇鞭菊在风中摇曳……

　　伫立在办公楼的楼顶，可以眺望远处的群山河流；俯瞰庭院，凉亭等景观尽收眼底。除此之外，设计公司还为该项目设计了指示牌、路灯以及门廊等设施。

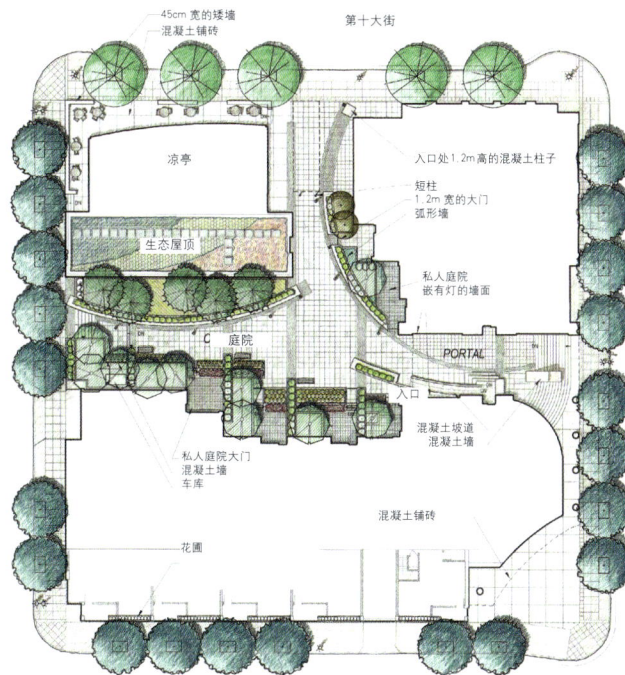

45cm 宽的矮墙
混凝土铺砖

第十大街

凉亭

生态屋顶

入口处1.2m高的混凝土柱子

短柱
1.2m 宽的大门
弧形墙

私人庭院
嵌有灯的墙面

庭院

PORTAL

入口

混凝土坡道
混凝土墙

私人庭院大门
混凝土墙
车库

混凝土铺砖

花圃

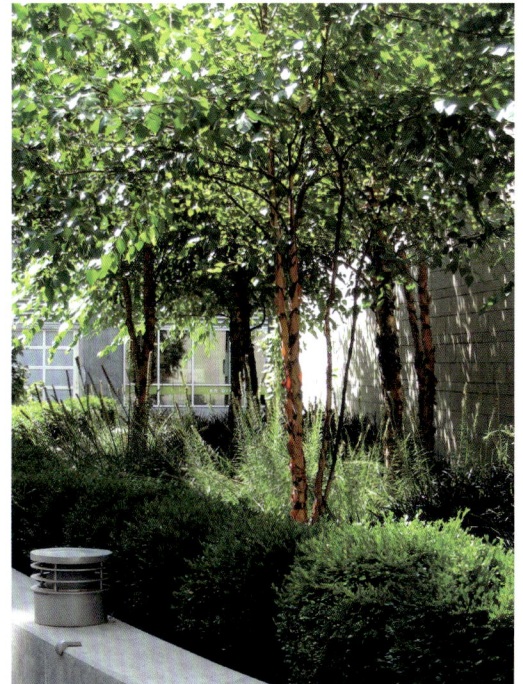

The Pinnacle courtyard is nestled between a single story retail pavilion and two wings of a residential condominium tower, 12 and 7 stories tall. From the condominium units, one has views over and beyond the cityscape to scenic mountains and below to the geometric courtyard landscape. This dynamically composed urban courtyard is built over underground parking and incorporates both intensive and extensive eco-roof technologies. Centered on the northern gateway to Portland Oregon's River District, KLA's design responds to the site being between the Willamette River walkway, the Fields Neighborhood Park and the River District's new urban renewal mixed use neighborhood.

Designed to accommodate a pedestrian passage through the site, grand arcing walls, custom lighting and planters guide people through the courtyard to an open breezeway that penetrates the building at the intersection of the wings. Varying heights of planter walls help to maximize the privacy of living units that also front on the courtyard space. The color and texture of plantings are choreographed to accent the buildings. Eastern Redbud trees, their early bronze foliage and magenta flowers compliment the subtle hues of purple black and red in the brick façade of the towers. The form and exfoliating bark of River Birch bring texture and interest to the tan color and smooth faced concrete masonry block walls of the pavilion building. Flowering spikes of Purple Liatris meander through mounds of Black Mondo Grass that rise above hedges of Boxwood to provide a bit of whimsy to the space.

The retail pavilion's roof is an extensive eco-roof planted with sweeping monoculture of different flowering sedums that mimic the arching pattern of the courtyard planters and paving, providing interest for the residents above. The design by Koch Landscape Architecture (KLA) also included custom signage, bollards and interior lobby furnishings.

办公区的清新景观 —— 瑞士奥普斯·楚格庭院

Refresh the Landscape in Business Area — Courtyard of Opus Zug, Switzerland

撰文 / 图片提供：Pedro F Marcelino　Planetage Landschaftsarchitekten　　翻译：沈翀

在城市规划这一背景的驱动下，设计师在奥普斯办公大楼的庭院内营造了人工景观，并使其成为办公大楼的核心。该项目的设计以水景和植栽为特色，与周围工业区的硬质铺装和夏季里干燥闷热的小巷形成对比，设计还包括许多新鲜的自然景观元素，如湖边的芦苇、湿地以及沼泽地。

天然的水族馆

水面的倒影和办公大楼表面反射的景色不仅丰富了庭院空间，而且与周围的建筑物相互融合。从办公大楼的高层向外眺望，庭院空间的景色一览无余，水景和植栽可以帮助缓解双眼的疲劳。与人等高的芦苇占据了整个庭院，并一直延伸到周边的步道上。站在紧贴水面的金属桥上能欣赏到更多水景和植栽，站在狭窄的金属步道上，人们同样可以看到流水和光线的反射。轻风在芦苇丛中低吟，为这里迷人的风景平添了一种特殊的音响效果。

独特的元素

该项目中的植栽除了作为富有创造力的设计元素以及生态系统的一部分，还能使人们观察并体会到其富有生命力的生长过程。

在冬季和早春，庭院主要以流水、沙砾和金属桥为特色，植栽并非设计的重点，而是以光线的闪烁、水中的倒影、与水亲近的感觉以及水天一色的体验为重点。宽广的水面影响着庭院内的"小气候"，提高了空气湿度并可以在夏季降温。芦苇、灯心草和鸢尾属植物高低错落，每年秋天都对其进行修剪，使其呈现出规整的形状。水景的重点则是水位稍深的区域内种植的色彩夺目的水生植物。

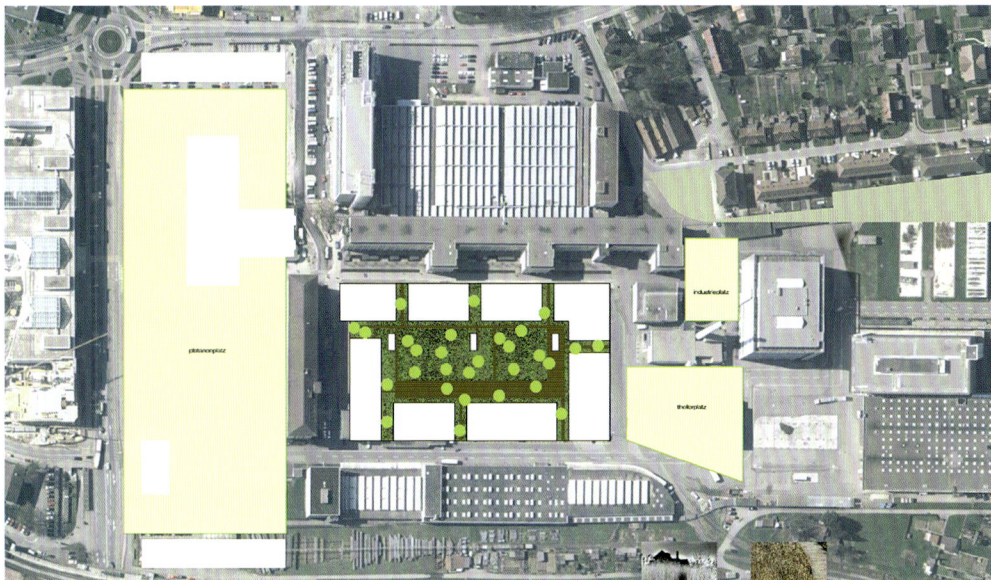

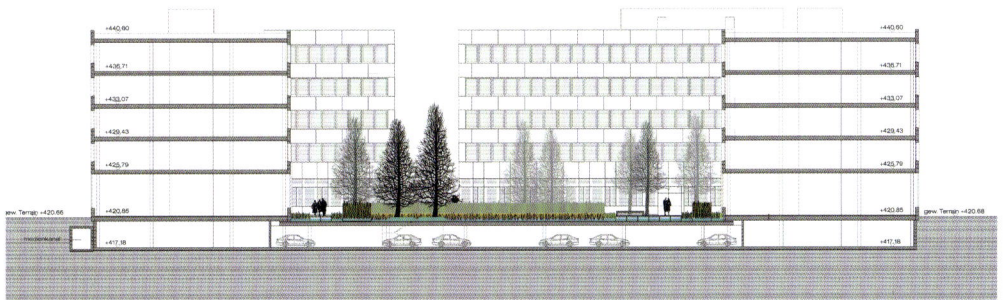

Hinweis:
Die Dammkrone muss unterhalb des Steges zu liegen kommen.
Stahlbauteile (Rost / Beleuchtung) dürfen den Damm, bzw. Dammkrone nicht berühren.

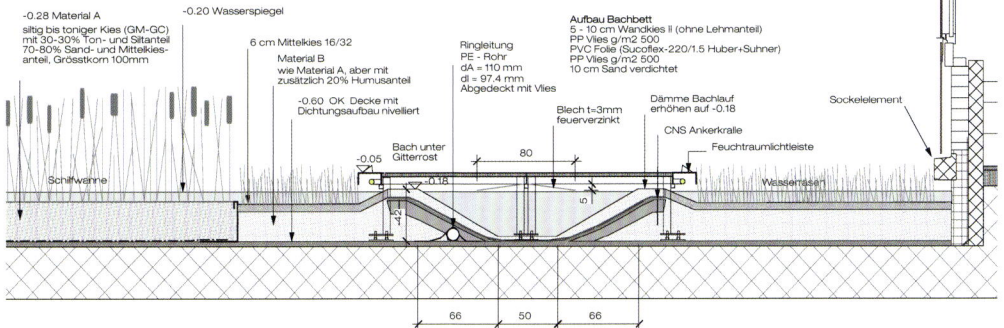

-0.28 Material A   -0.20 Wasserspiegel

siltig bis toniger Kies (GM-GC)
mit 30-30% Ton- und Siltanteil
70-80% Sand- und Mittelkies-
anteil, Grösstkorn 100mm

6 cm Mittelkies 16/32

Material B
wie Material A, aber mit
zusätzlich 20% Humusanteil

-0.60 OK Decke mit
Dichtungsaufbau nivelliert

Ringleitung
PE - Rohr
dA = 110 mm
dI = 97.4 mm
Abgedeckt mit Vlies

Aufbau Bachbett
5 - 10 cm Wandkies II (ohne Lehmanteil)
PP Vlies g/m2 500
PVC Folie (Sucoflex-220/1.5 Huber+Suhner)
PP Vlies g/m2 500
10 cm Sand verdichtet

Dämme Bachlauf
erhöhen auf -0.18

Blech t=3mm
feuerverzinkt

CNS Ankerkralle

Feuchtraumlichtleiste

Sockelelement

Bach unter
Gitterrost

-0.05

-0.18

Schilfwanne

Wasserrasen

66   50   66

80

The artificial landscape created in the courtyard of the opus office buildings is to be understood as an intervention in the urban context. The courtyard—like the core of a babuschka of office buildings—is characterized by water and vegetation and is being created as a soft contrast to the hard surfaces of the surrounding industrial area with its narrow lanes which get very dry and hot in summer. The landscape scenery consists of various pristine natural elements of landscapes such as the reed alongside lakeshores or the swamps of the everglades in Florida, whereas the alienated context is of importance.

An Accessible Aquarium

The reflections on the water surfaces as well as on the facades of the buildings dissolve the dimensions of the courtyard. The landscape scenery is duplicated and melts together with the surrounding facades. From above—looking out of the office windows—the dimensions of the courtyard can be measured with a single glance. The image of water and vegetation offers relaxing views and is a benefaction for tired eyes looking into monitor screens all

day. At pedestrian level the reed, which appears to be like a filling of the courtyard, oozes out to the surrounding streets and pedestrian areas. Passers-by are attracted to discover the unknown and possibly to remain for a while.

Even more intensively water and vegetation are being experienced from the metal bridges which are lay close to the water surface. The flowing of the water and the reflection of light are getting visible from these metal catwalks.The rustling of the wind in the reed is creating the acoustic backdrop of this attractive and diverse scenery.

Growths and Development as a Distinctive Element

Besides the use of plants as an creative element and part of an ecological system in this project plants can be experienced in their specific dynamics of growth.

In winter and early spring the courtyard is mainly characterized by water, gravel and the metal bridges whereas plants are not of importance to the design —important is just the play of light and reflections, the strong feeling of being close to the water, the experience of wideness in combination with the sky. The large size of the water surface affects the microclimate of the courtyard in a positive way, it enhances the humidity of the air and cools temperatures down in summer.Reed (consisting of Typha minima, Typha angustifolia, Phragmites australis), rushes and irises (Juncus effusus, Juncus ensifolius, Schoenoplectus lacustris, Iris pseudacorus) are creating a carpet of planting with changing heights. Each autumn the planting is being cut down and gives view to the trunks of the taxodiums which sit on top of small gravel islands. Centrepiece of the waterscape are the slightly deeper areas with colourful and finer waterplants such as waterlilies and Menyantes trifoliata.

现代地景的演绎艺术 —— 苏州工业园区行政中心

The Art of Transformation Landscape — Suzhou Industrial Park Administration Center

撰文：Elizabeth Shreeve　Chih-Wei Lin　　图片提供：SWA Group Sausalito　Tom Fox　John Loomis　Hui-Li Lee

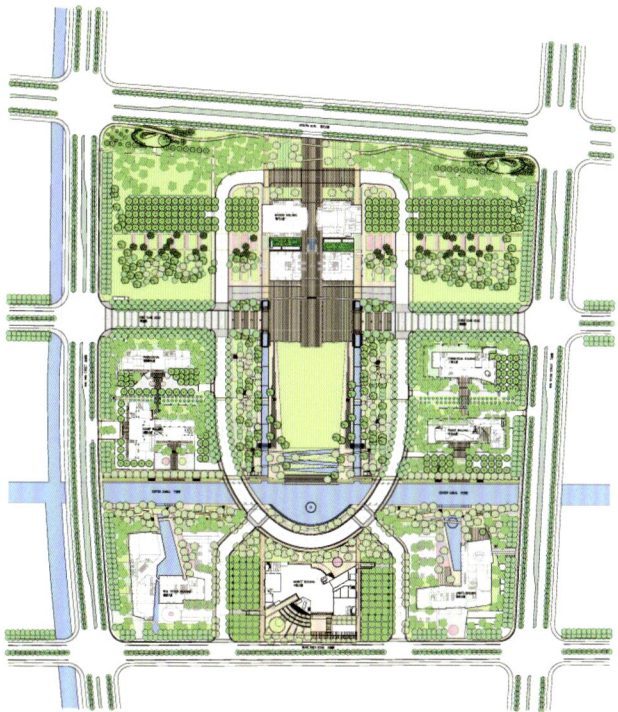

苏州工业园区位于金鸡湖的东侧，在新式工业发展规划的架构下，昔日驰名的"水乡园林之都"的风貌如今已被大型格状街区、规划工整的厂房和高速公路系统所取代。行政中心坐落在园区的入口处，与两侧的旧城区隔湖相望。

该项目由8座建筑组成，分别是北面的行政大楼、西面的检察院和法院、东面的工商大楼和公安局、南面的市场营销、房地产和置地大楼。全区跨越了9个街区，总占地面积约283 000 ㎡，是一处集政府机关日常办公以及市民活动于一体的大型公共场所。设计师在该项目中面临的主要问题在于如何通过全区景观的规划，有效地整合这8座散布的建筑，并在户外空间中创造出既代表苏州传统文化的景观，又兼具现代气息的地景风貌。

在总体配置上，景观设计方案以多排并植的香樟林为骨架，与人行步道系统相结合，将建筑群体整合起来，形成独特的行车体验；各区域的停车场也通过刻意的布局与大片的树林融为一体；一排排的树木和地面的铺装设计也都反映出建筑现代简约的立面风格，加强苏州新行政中心的整体性。

方案中的主要空间包括行政大楼前的迎宾道、集会广场、中央草坪、交织广场、东西两侧的林阴花园和水道的中央公园、以及沿着南面渠道设计的带状公园。在这一系列的开放空间里，水——苏州都市的灵魂元素，是统一整体设计的基调，设计师通过现代设计手法将这个传统元素重新演绎转化，创造出不同的空间类型和功能。

通往园区中心的东西迎宾道连接着周边的道路和内部环路。利用特殊的石材铺装以及两旁3排交错种植的大树，将入口界定为大型公共空间。集会广场延续了行政大楼的风格，结合切割石材的铺装，在建筑与景观的垂直空间与水平空间的转换过程中营造出强烈的视觉延伸效果。大型的市民集会与公众活动都能够在这个广场上进行。进入集会广场后，人们的视线就会立刻被开放的中央草坪与两侧的线形水道所吸引。集会广场与水道会合的地方设有3m高的瀑布，广场的北面更搭配一组矩阵式喷泉在空中起舞。水道巧妙地与太湖石景相结合，通过高低有序、疏密有致的精心布局，赋予江南庭园中的典型石林独特的现代

风格，成为园区内的重点标志景观。水道向南延伸，在端点的设计上巧妙地暗示着其与贯穿城市的渠道合二为一。

该水道缓缓向南抬高，形成了环抱中央草坪的花岗岩台地，强化了中央向南延伸的视觉走廊。外侧的林阴花园则是由一系列的波形雕塑与线形混植的水杉、银杏、榉树和合欢林组成。波形雕塑试图撷取水的流动性，在花园中创造活泼律动的趣味性，并且为在其间休憩的人们制造自然优美的屏障。中央草坪的南端是交织广场，错落交织着楔形的映面池、草坪，步道穿梭其间。这里适用于非正式的活动，如户外艺术展览或观赏渠道中央弧形的水舞喷泉等。

保留下来的城市运河的南岸是一座带状公园，配有高低错落的观景平台，可以眺望水流；垂柳和枫树提供了四季变化的色彩。不同高度的平台之间由台阶和无障碍斜坡道连通，最低处临水的漫步道更可以使行人在环路桥下畅行无阻。

该项目完工至今已经成为市民日常生活中不可缺少的活动场所，重新联结了苏州现代与传统的文化景观；而多元的空间不只体现了现代地景的精神，更具备了具有文化深度的美感。

Suzhou Industrial Park is located at the east of Jingji Lake. Under the new development of industrial park, the well-known landscape of "Canal Garden City" has been replaced by large scale blocks, efficiently designed ware houses and high-speed roadway system. The Administration Center is sited at the entry of the Industrial Park, facing old town on the west with the lake in-between.

Suzhou Industrial Park Administration Center is composed of eight municipal entities—Administration Building at the north, The Prosecution and Court to the west, Commerce and Police office to the east, Marketing, Real Estate, and Land Buildings to the south. The site occupies 9 blocks with a total area of 28.3 hectares and functions as a place for government services and civic gatherings. The main task for SWA on the project is to integrate the eight buildings through landscape master plan and create landscape that depict the beauty of traditional Suzhou culture with contemporary expression.

The schematic landscape plan utilizes formally planted matched Camphor trees to form a green structure throughout the site. Coordinating with pedestrian circulation system, this green structure strongly integrates the building entities, and creates a unique driving experience for the street. Ground parking areas shown in the Master Plan are incorporated within the larger bosques of trees. The tree rows in the bosque and the ground patterns are all designed to reflect the minimal style elevation/facade of the building and help consolidates the integrity of this new administration center.

This great space is composed of ceremonial entry to the Administration Building, a grand plaza, central lawn, weaving plaza, central park with elegant linear water feature and flanking tree gardens, and a canal front park. Water—the key Suzhou image is the fundamental element used to unify the designed spaces. It is being transformed and re-endowed with diverse functionality and presence.

The experience of entering the site and central space is heightened by special paving on the east and west roads that connect the surrounding boulevards with the internal loop drive. Framed by triple rows of trees and paved with accent stone bands, these roads announce through sight and sound, the entry as a grand civic space. The Grand Plaza

is contemporary in expression paved with cut stone in a pattern that reflects the articulation of the Administration Building façade. It is designed to accommodate large civic gatherings and public events. Once within the Great Plaza the eye is drawn toward the open central lawn and the linear water feature that extends southward toward the canal. Where the plaza and water feature meet, a pair of 3m high sound-box water fall and a grid of fountain jets dance in the air and spill over into the water feature with an active and refreshing sound. The linear water feature is formed by a sequence of stepping pools with special layout field of Taihu stone. This feature has transformed the traditional garden material into a modern expression and becomes the landmark focus of the Center. At the canal end of the linear water feature the water disappears into a square trough creating the waterfall sound for the weaving plaza.

This striking water feature ascents toward the south, and forming two flanking granite berms that rise up from the water feature and frame the central lawn, creating forced perspectives and promontories viewing to canal. The other major element of the central park is the two flanking tree gardens. These multiple rows of shade trees are interspersed with a series of sculptural grass berms and small plazas. They create a metaphor of waves emanating from the fountain at the Grand Plaza down the canal. The spaces are shaded sitting areas, small gathering areas and informal recreation places. The sculptured berms screen out sounds from the adjacent loop road. To the south, a weaving plaza appears with wedge shape pools and lawn and pedestrian path weaving through. The weaving plaza forms the south edge of the central park and is suitable for informal activities such as outdoor art exhibits or small amphitheater for watching the water dancing at the center canal.

On the south side of the preserved urban canal is a linear park with a series of terraces affording views toward the canal and steps and ramps descending to a lower level walkway along the canal. This lower level allows for pedestrian continuity under the loop road along the south edge of the canal.

The Center has become one of the most important civic spaces in Suzhou. It successfully bridges the traditional and modern cultural landscape, providing virtues which not only reside in modern spirit but profundity of Suzhou history.

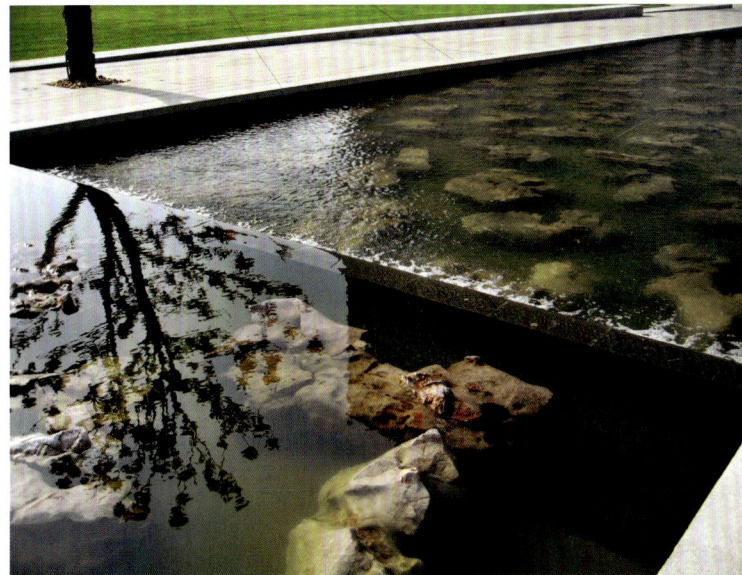

回归与延续 —— 美的总部大楼景观设计

Return and Development — Landscape Design for Midea Group Headquarter

撰文 / 图片提供：广州土人景观顾问有限公司

1 水景
2 景观构筑物
3 休闲广场
4 地下车库出入口
5 湿地景观
6 南广场
7 北广场
8 旗杆及早喷水景
9 主题雕塑
10 芭蕉
11 采光井

总平面图

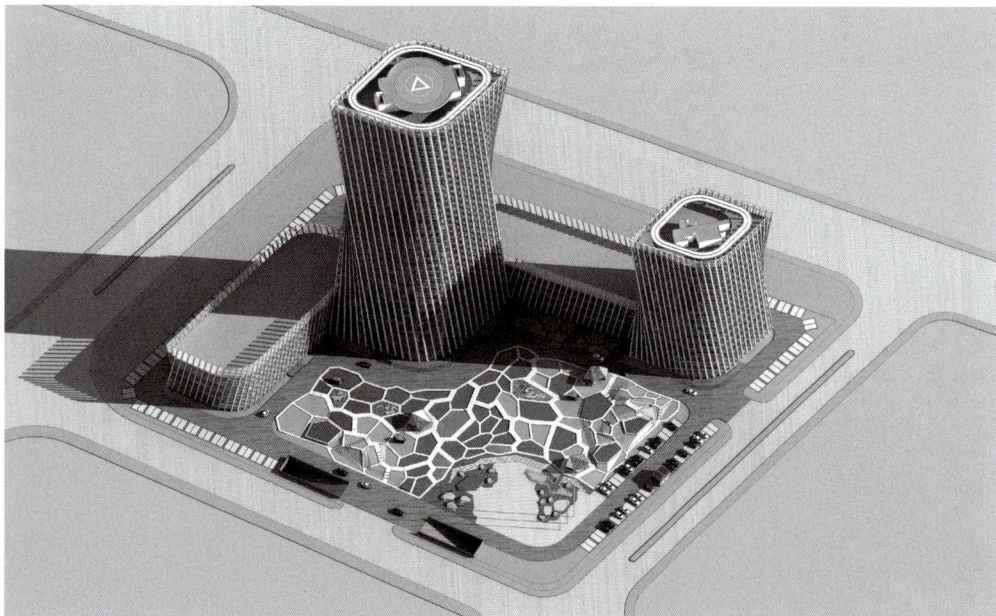

轴测图

该项目处于顺德北新城区的住宅区与工业区的包围之中，占地面积为 23000m²，总部大楼高 128m、共 31 层，是目前顺德最高的地标建筑；并成为了"2010 年第六届欧洲景观双年展——中国当代景观展（水之中国）"入选的 15 个参展作品之一。

问题和矛盾

桑基鱼塘，其产生源于广府境内河流纵横、地势低洼，涝水容易为患。人们根据本地自然环境的特点，一则修堤坝治洪水；另则挖洼地为塘、降低水位，基高塘低，基种作物、塘养鱼虾，形成良性循环的农业生态环境，创造出独特的农业生产方式——桑基鱼塘。这种独具珠江三角洲特色的农业景观尤以广东省内的南海、番禺和顺德居多。

桑基鱼塘的"基"与"塘"构成的网状肌理无论是从高空俯瞰还是身处其中，都给人留下异常深刻的印象。其独具特色的地理环境、地域历史和地域文化，造就了独特的珠三角景观——"密布的河网，成片的基塘，繁茂淳郁的花木，杂糅着低吟浅唱的虫鸟"，这是在珠三角地区、尤其是顺德，每个人心中都能唤起的亚热带景观想像。

然而，这些年不争的事实是农业文明时代的景观地图早已被高速的城市化进程所改变。熟识的乡土图景、温润的乡情已逐渐被致密紧凑的城中村、大而不当的硬质广场和冷漠的邻里关系所替代。经历本土农业文明的文化语境到后工业文明文化语境的转换，独具珠江三角洲农业特色的桑基鱼塘已逐渐变成了关于过往的回忆。

设计对策

该项目通过现代景观语言回应了中国岭南大地景观"桑基鱼塘"和珠三角水网，在高速城市化的当下回归乡土景观形式与本土美感意境。

阡陌交通的栈桥和道路将场地分割成大小不等、

形态各异的几何体——或下沉为水景，或上浮为种植乡土林木的小丘，或为区域小广场（庭院），或为地下室采光天井。并在其中点缀以乡土材料建造的现代景观构筑物，以形态和乡土材料的组合来解决凸起的若干地下室采光天井的视觉问题，并贯穿、延续地域景观。用栈桥、道路、水景与庭院等实际功能体块勾勒出"桑基鱼塘"的网状肌理，让人体验到的不仅是肌理间生动丰富的功能联系，还有亲切舒缓的基塘肌理带给人的仿佛当年人对土地的归属感。

水体应用

水景在该项目中被分成生态湿地以及地下室采光天井上的薄水，其重点不在于展现水景的不同形式，也不在于水景带来的若干亲水活动，而是在于对区域文化、生活及当地自然环境关系的尊重。

地下室采光天井上的薄水以景观的手法解决了众多地下室天井的采光问题，并与生态湿地一样作为雨水、污水收集处理系统的一部分。把屋面和露天雨水收集、处理、蓄积在景观水池之中，将产生的中水和污水全部回收，通过生态湿地进行生物降解处理，重新用于绿化灌溉和补充景观水池水量，无需用饮用水作为景观用水。

水景设计也解释了人们面对环境状况所做出的选择——延续和存在。

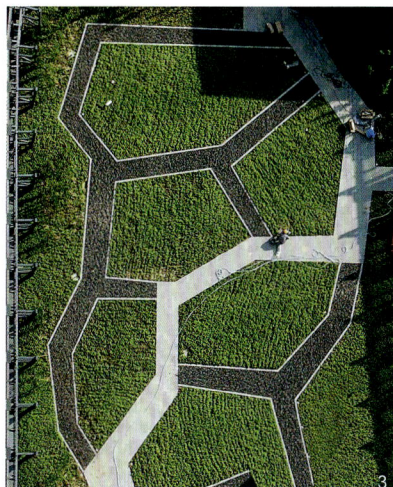

1　全景图

2、3　屋顶花园

Media Group(Shunde famous Chinese home appliance leading enterprises) new headquarters building is located in Shunde city surrounded by residential and industrial area, the headquarters building of 31 storey, 128 meters, is the tallest landmark in Shunde. The project covers an area of 23,000 square meters. The project is one of the selected works of 15 exhibitor in the "2010 Sixth European Landscape Biennial—Contemporary Landscape Exhibition (Water China)"

Problems and Contradiction

"Mulberry-dyke & fish-pond", its territorial landform is originated from rivers networks in Guangdong Province, the local people according to the characteristics of natural environment, reduced flood influence by dyke which can cultivate crops and dig fish ponds for water retention. This unique characteristic agriculture landscape is prevailing in the Pearl River delta of Guangdong province, especially in districts of Shunde, Panyu and Nanhai.

Mulberry-dyke & fish-pond constitutes a mesh-like texture whether in overlooking aloft or standing in, and anyone would be branded by deep impression. Its unique geographical environment and regional history and culture bring up an exclusive Pearl River delta landscape. Clouds of the rivers and ponds, lush flowers and trees, mingled with humming birds and insects singing, which is the Pearl River Delta region, particularly in Shunde, in every heart can

土地肌理的延伸

evoke subtropical landscape imagination.

However, these years indisputable fact is that the landscape map of agricultural civilization had already been changed by high-speed urbanization process. Familiar countryside pictures, is gradually replaced by compact and dense urban villages, improper squares and distant neighborhood. Through the local cultural context of agricultural civilization to industrial civilization, after the conversion, unique characteristics of the Pearl River Delta agriculture—"Mulberry-dyke & fish-pond" is becoming past memories.

Design Strategies

Media group headquarter landscape design used modern design languages respond to southern China "Mulberry-dyke & fish-pond" and the Pearl River Delta water network, present in the rapid urbanization of the rural landscape and return to the local beauty of artistic conception.

Crossroad trestle and road traffic will divide land into different size and shapes of geometry—or sink to water features, or mound for planting native trees, or a small square area (courtyard), or the basement skylight. And dotted with modern landscape construction made from local materials and it solves the basement skylight problem by form and combination of local materials, continuing execution of regional landscape. With the trestle, roads, water features and garden features, actual physical block outline "mulberry-dyke & fish-pond" mesh texture, not only people to experience the texture by contacting between the vivid and rich features, but bring sense of land belonging in old time.

Application of Water Feature

Water features in the site was divided into the wetlands and the thin water above basement skylight, the concept is not to reproduce the different forms of water features, and not about water activities, but respect the region culture, life the relationship between the local natural environments.

Wetland and thin water above basement skylight are designed for rain, sewage water collection and as part of water treatment systems. The rainwater will harvest, accumulate, and collect in the pond will have the full reclaim by biological degradation through the wetlands treatment, and use for the green irrigation and supply pond water, do not use drink water as landscape use.

Waterscape design also explained the state of the environment we are facing today, the choices we make - how to continue and exist.

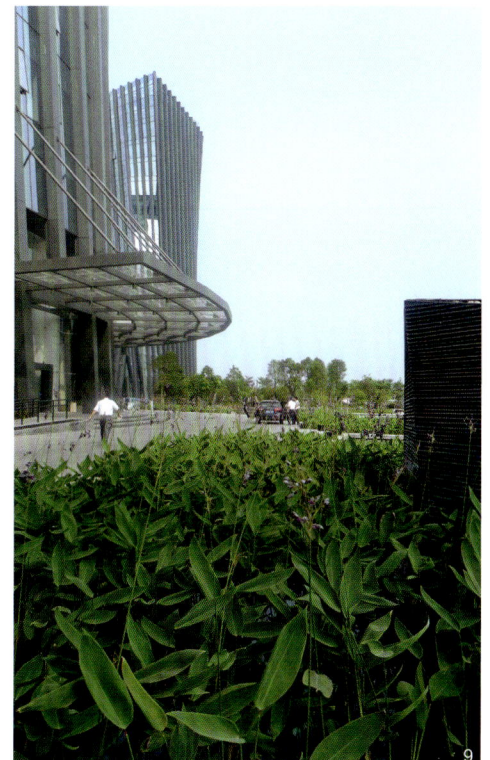

乌得勒支的"墨水池"

"Inktpot" Utrecht

撰文：Martin Knuijt　　图片提供：Ben ter Mull　　翻译：王玲

紧邻乌得勒支中央火车站的荷铁前总部"De Inktpot"（又称"墨水池"）是荷兰最大的砖造建筑。"墨水池"不仅仅是它的一个别称，还蕴涵了人们对建筑的倾慕之情——整座建筑气势恢弘、精雕细琢、典雅精美。在该项目的设计过程中，设计师大胆创新，不仅使庭院的户外空间更具现代感，而且使其与沉稳宏大的砖造建筑浑然天成。

目前以实用功能为主的庭院都无法体现出建筑作为铁路网络物流中心的功能和铁路运输组织的重要性。在设计方案中，设计师将表面的矛盾和限制巧妙地运用，形成一个富于创造力的庭院。无论是各种场景的布置还是其幕后的设计，均兼顾了多重用途。当前的实用型设计与诸多的娱乐型设计相辅相成，不仅满足了各种功能需求，还别有一番情趣——人们沐浴着夏日的阳光，沉醉于户外的一片绿意盎然中。设计的第三个功能是呈现"满眼的翠色欲滴"的视觉效果——设计师利用非常有限的建筑承载力，打造出充满生机的绿色屋顶。内庭院的建筑屋顶正好满足了高层办公区用户的功能需求。

富于动感的设计

庭院设计灵活多变、富于动感；既注重实用性又充满情趣，二者的巧妙结合从钢筋混凝土地面开始就被淋漓尽致地展现出来。景观小品包括轨道上可自由移动的植栽容器以及随意组合的庭院坐椅——当需要更多坐席或私密空间的时候，这些景观小品就可以进行相应的变换。如升降车需要更多空间时，只需将所

有的景观小品简单地推到一边即可。此外，可自由移
动的坐椅和遮阳伞也增加了庭院空间的灵活性。

　　庭院设计在有限的空间和采光条件下融合了不同
的功能需求——它巧妙地"捕捉光线"，使光影在明
亮的空间中舞动。地面采用浅色调，并选择具有反光
性的材料。植被的纹理及其特有的单薄感勾绘出一幅
淡然的景象——昏暗背景下的精美树叶和细弱茎干彰
显出一片旷远深幽。

聚会庆典

　　每逢聚会庆典，庭院的气氛都会达到高潮。无论

是美妙的花园派对还是在圣诞树装点下的聚会，都会
令人流连忘返。于是，顷刻间庭院摇身一变，圣诞树
和其他的装饰品将其装点一新——还安装了大量的电
源插头，如在圣诞树的底座上、笔记本电脑连接线上
或遮阳伞底座上，而所有拉线的终端都固定在墙面上。

　　该项目的设计仿佛一张活动的日历。当然它的
实际用途还要取决于使用者的自身需求。庭院并不仅
仅是人们在午餐时间休息片刻的场所，它也可以是摆
着华丽餐桌的复活节用餐场所或炎炎夏日里的啤酒花
园，还可以利用长桌使内部空间变得更加舒适，或者
在秋日里供人们聆听悠扬的弦乐四重奏。

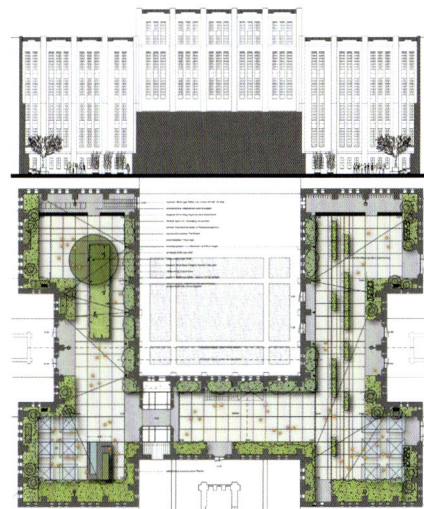

Just next to Utrecht Central Station you find "De Inktpot" (the Inkwell), the largest brick building in the Netherlands. More than just a colloquialism or nickname, "De Inktpot" has become a term of endearment for the building. The building is of an imposing size, but with a refinement in the details and finishing touches. For the design of the patio, OKRA was asked to develop a change programme to make the outdoor space of the patio more contemporary while keeping it in harmony with the static monumentality of the building.

Nothing in the present, utilitarian-designed patio reflects a design for the building's function as logistic centre of the railway network or the significance of the railway organization. In the plan developed, the seeming contradictions and limitations are utilized to create a patio that speaks to the imagination. The design sets the scene for various uses, both the stage and its sets and the world behind the scenes. The current "utilitarian" use is supplemented with more of a "use for enjoyment," which is nonetheless just as functional, but of another order: sitting in the summer sun, getting outside together and having a break meeting while enjoying the green. The third function is that of "green for the eye." The roof of the building in the inner courtyard can fulfill this function for the users of the upper floor offices adjacent to the interior space. It becomes a green roof, utilizing the very limited load-bearing capacity of the building.

Dynamics in design

The heavy utilitarian program and the use of the patio for pleasure are joined together by starting with the modular stelcon plate floor and going from there. The dynamics lie mainly in the design. The furnishings consist of mobile planters on rails and loose patio chairs. Any time more seating is required or there is a need for more seclusion, the furnishings are moved accordingly. If the hoist vehicle requires more space, everything can be simply pushed aside. In addition, extra objects, such as a patio floor with moveable chairs parasols increase the flexibility of the space.

A variety of uses come together in an area of limited space and limited light. The area is designed to "capture light," so that gloom does not dominate and instead, a strong image is created. The floor becomes light, and materials that reflect some light are chosen. The texture of the greenery and a certain degree of thinness in the greenery set the tone for the experience: Delicate foliage and thin stalks against a dark background create depth.

Festivities

Moments of festivities are temporary high points, which mean a lot for the occupation of the patio. Everyone knows the sort of thing, a cracking garden party or a special meeting by the Christmas tree outside. For a short time, the patio becomes something else. It is dressed up with streamers, a Christmas tree or other attributes. A number of plug-ins are installed, such as a stand for the Christmas tree, connections for laptops or feet for parasols. The walls are fitted with anchors for bracing wires.

As a starting point for the use of the patio, OKRA presented an activity calendar. The actual use will, of course, be determined by the users themselves. With an active canteen manager or personnel association, there will be more than just popping out to smoke or sitting outside for a while during lunch. The redesigned patio is ready for an Easter brunch with opulent table settings just before Easter, the beer garden in high summer, using long tables to transform the interior space into a pleasant attraction, or the string quartet in the fall.